"I HAVE COME A LONG WAY . . ."

The incredibly beautiful woman in translucent garb beckoned to Howard. He felt a tremendous emanation of warmth and love as he moved toward her. Much of what she told him — of other planets, of spaceships — was mysterious to him, but not her ultimate, thrilling message:

"We are contacting our own."

For this was only the first of many meetings Howard Menger had with the uncanny visitors from beyond our skies — and in this extraordinary book, Menger shares his unique adventures with the reader — including actual trips in flying saucers!

FROM OUTER SPACE

(original title: *From Outer Space to You*)

HOWARD MENGER

PYRAMID BOOKS • NEW YORK

The Last of Atlantis

The Atlantian stood in the blazoned night
Within the disk so burnished bright,
Poised to launch him into unknown space
At speeds unknown in this earthly pace.

This man was the best the race could find:
Sound of body and sound of mind;
The highest evolved of all their men
To command this ship and the crew of ten.

It was he who'd race at the speed of light,
The first to see the galactic might
Of star islands only known by few
And not yet seen in earthly view.

A thousands years they'd said he'd be
Streaking through the galactic sea
To a target 500 light years hence
In a state of time held in suspense.

With a mighty surge the disk took flight
Upward, upward into endless night,
Lost soon in the starry heavens black
As this man cast one long glance back.

Years to decades to centuries became
Man learned to hurt, to kill, and maim,
Slipping backward from the heights attained
As the light of love grew dim and waned.

As the earth returned to a barbaric age
A new life cycle turned its page.
The earths' crust split and spewed up flame,
And the boiling sea sealed Atlantis' shame.

* * * * * * * * * *

Ten centuries passed and the ship returned
To an earth now degenerate and burned,
Not fit to sustain the life of few
Who'd lived the Creator's love and knew.

They hovered for months about our earth,
Vainly seeking a new hope's birth,
Saw naught but ego, hate and pain—
Man against man for paltry gain.

With heavy heart and saddened eye,
This man of space soon cast the die,
Turned the ship outward to endless space
To find a home to sustain his race.

In the spiralling star dust lies a far-off land
Where even now they'll lend a hand
To raise us from our knees to see
The errors that cast them beneath the sea.

C. B. Brailey

MEET THE CREW

To all those who have been directly or indirectly instrumental in assisting him to bring this message of hope and enlightenment from our brothers of other planets, the author wishes to extend his humble and sincere thanks. He has space to name only a few of them:

"Long John" Nebel of WOR, New York, who gave so generously of his time so that the author might share the message of the Visitors.

Frank Ford Show, WPEN, Philadelphia, Pa.

Gene Crain's "Wonderful Town" WCAU, Philadelphia, Pa.

Vern Craig's "Open Mike" Allentown, Pa.

Jack Paar Show, N.B.C.

Dave Garroway's "Today" Show, N.B.C.

And many other radio and TV shows.

Washington Star, Washington, N.J.

The Daily Record, East Stroudsburg, Pa.

Newark Star Ledger, Newark, N.J.

Philadelphia Enquirer, Philadelphia, Pa.

The Indianapolis Star, Indianapolis, Ind.

The Chicago American, Chicago, Ill.

LIFE Magazine, New York.

ARGOSY, New York.

Bucks County Traveler, Bucks County, Pa.

Many other publications and periodicals, among them George Van Tassel's *Proceedings,* Yucca Valley, Calif., and *Flying Saucer Review,* London.

Calvin C. Girvin, for the dust jacket painting.

Don Leigh McCulty, for editorial assistance.

August C. Roberts, for photographic consultation.

And, of course, the space people themselves.

For free current information about Unidentified Flying Objects, write to Spacecraft Research, Box 1941, Clarksburg, West Virginia 26301. To reach the author, write to Howard Menger, in care of the above address.

BRIEFING

It is evident that we are past the threshold of a new era, not only in scientific advancements, cultural growths, and the ethical and social development of all peoples, but in two other major areas of life: the geophysical, which has to do with the physical changes going on and about to take place in, on and above our Planet Earth; the other is the spiritual renaissance of all people on the planet.

The geophysical changes have already been predicted by our own scientists, with their announcements that the axial motion of the earth is 23 degrees off; with the piling up of ice on the South Pole, affecting sea, air and land masses; and the cyclic changes, which seem to take place periodically.

The spiritual growth of the people is, of course, the universal desire for freedom and liberty and the worship of One Creator, along with the brotherhood of man and the Fatherhood of God.

If, due to our experiments with atomic and thermonuclear explosives, we are dealing nature a bad blow to an already weakened condition, two things must be prevented: atomic and hydrogen blasts—and general warfare.

Let us assume that there is on this planet a group of scientifically minded and spiritually dedicated men and women who are working to accomplish this great task. And, let us further assume that they have already established contacts with equally dedicated people of other planets. To continue their work and remain effective, they must of necessity remain behind the scenes. However, they can, in the interest of humanity in general, send out hints as to what will take place in the near future. Perhaps they send out scouts to make personal contacts for the specific reason of determining the reactions of every-day people. Perhaps it is done as a "smoke screen" to temporarily keep secret the real work which is going on until such time that the people are prepared to meet this new era with many changes it will bring.

There are perhaps, many bases of operation already established on, under and above this earth which prepare to meet this coming challenge. The increased sightings of spacecraft give evidence to the hidden activities. And what are these "signs in the sky" so many people are seeing, and why do descriptions vary, and how can the craft seem to appear and disappear? Perhaps all this can be explained by simple laws we have not yet learned about. Or perhaps they are merely projections, designed to divert or attract attention, depending upon the circumstances involved.

Then there are the personal contact stories, some of which are authentic, and which have been established for study purposes and for keeping alive a story which must eventually be brought before all people. If given in small doses, the general acceptance will be made over a period of time, and will take place almost naturally.

Let us imagine, then, that this great work is being carried on by a universal group of men and women with contacts in every government in the world (world leaders included) and in every walk of life. And this group, I would like to believe, is dedicated to saving mankind and this Planet Earth, so that we too can step out into the Universe and travel to our neighboring worlds.

Marla

PRESENTING THE PILOT

Introductory speech made by Cortland Hastings at the Pythian Temple, New York City, December 16, 1956.

My name is Cortland Hastings. My function is to introduce to you Howard Menger. But first, let us adjust our perspective—look at ourselves and the earth in relation to the universe.

Astronomy is the science of studying celestial bodies: their positions, sizes, motions, constitutions or compositions, their mutual relationships, their histories, their destinies. Astronomy treats of the earth only in relation to the celestial bodies.

The first question is, what is the age of the universe? The answer: the universe is ageless. Some orthodox religions teach that the earth is but a few thousand years old; however through the study of uranium, with its progressive half-life reduction every five billion years (that is, every five billion years it loses 50 per cent of its strength), the earth has now been found to be some three or three and one-quarter billion years old. And there is no reason to suspect that the earth represents the age of the universe.

What about the extent of the universe? Man, with his finite terms, cannot truly express the Infinite. But he can and must try.

The great 200-inch telescope at Mt. Palomar, California, has photographed galaxies about one billion light years away. Now, as we know, a light year is a measure of distance, not of time. It represents the distance which light traveling at 186,000 miles per second will traverse in one year. In miles, then, a light year is five trillion, eight hundred and seventy-five billion miles. So if you took out a pad and multiplied that by a billion, you would come up with a round figure in miles of six with 18 zeros following it, the distance of these farthest galaxies yet photographed. And you should have a large pad, because that is six with 18 zeros following it.

But this is, by no means, the boundary of the universe; it is simply the farthest distance ever seen by man's instruments —remarkable as these instruments are, or seem to be.

The astronomer has many times been called an atheist. This is largely untrue. On the contrary, he is usually a very reverent man, because he can see, better than anyone else on Earth, the marvelous workings of the universe. The astronomer, with his scientific training, can also, more than we, appreciate the plan and order of the universe. He knows this law and order of the sky is not accidental, that it did not just happen. Even more than we, the astronomer can realize that this vast universe is the result and plan of Our Source— a Source of inconceivable intelligence and incomprehensible power.

It has been estimated that there are well more than one hundred million galaxies, each containing many billions of stars and planets. Our own galaxy, of which our solar system is but a small part, is the Milky Way. The Milky Way Galaxy is estimated at more than six hundred quadrillion miles in width. We do not know how long it is, or rather the circumference of its spiral.

Our solar system, as we know it, is composed of the sun and nine planets, of which the earth is one—and one of the smaller ones at that. The extent of our solar system is three billion, seven million miles. Now our sun is but a very ordinary star in our Milky Way galaxy. Many stars are a hundred times larger and brighter than our sun. A star, by the way, is defined as an illumined body giving off its own light, contrasted to a planet which simply reflects the light of a star, in the case of our system, the sun.

The sun is the source of light and life on the earth. At this point we can well ask, "From whence does the Sun get its light?" The answer is simply: from Our Source—the Source of our Universe. Now all space is not a void. It is filled with vibrations, rays, oscillations of many and energy—the initial oscillation from the Source of All. There is no place in this vast universe where Intelligence and Energy are not. So, space is not a void. It is filled with vibrations, rays, oscillations of many kinds.

There are electromagnetic rays both above and below the commercial radio band. The radio utilizes a band from five hundred thousand to a million and a half vibrations per second. Television utilizes frequencies higher than that: from 44 up to 108 million vibrations per second. Above that come the calorific, or heat rays, which range up to 35 trillion vibrations per second. Even above that are the chromatic or color rays, ranging from 430 trillion vibrations per second, which is the color red, to 700 trillion, representing purple. Then there are actinic rays, which perform chemical changes in our plant and animal life upon the earth, and which give life to the earth. These also emanate from the sun, and their frequencies are 600 trillion vibrations per second and higher. Above these we find the X-rays, or roentgen rays; and far above those the radium or gamma rays. And so we proceed into even another kind of infinity: scientists feel that above those frequencies they can now measure surely exist others, and others.

But let us get back to our solar system, because that is what we are going to be more directly concerned with. Our sun is 864,000 miles in diameter, and is spinning through space at 720,000 miles an hour. That is fantastic isn't it? How many people do you hear saying, "Oh, we never go anywhere!"? But at that very moment they are traveling at the rate of 720,000 miles an hour, because the sun is pulling along with it the entire solar system. We on earth are not sensitive to this speed because our atmosphere envelope is being pulled along at the same rate of speed.

Now the earth itself is a little less than 8,000 miles in diameter, and is located 93 million miles from the sun. Between the earth and the sun are two planets. The first, from the sun outward, is Mercury, one of the smallest planets. It is only 3,000 miles in diameter and is located 36 million miles from the sun. Next comes Venus, which is a little smaller than the earth—7,600 miles in diameter and located 67 million miles from the sun. Then comes the earth—7,920 miles in diameter, if you want the exact figure, and 93 million miles from the sun. Now, we go away from the sun to the rest of the planets. Mars, the next one, is 4,200 miles in diameter and 141 million miles from the sun. Now a big jump, through an area in space in which are spinning many thousands of asteroids and planetoids, which many an astronomer believes may be the result of a former shattered planet, which legends have told us might have been called Lucifer.

We now come to the biggest planet in the solar system: Jupiter, 87,000 miles in diameter and 489 million miles from the sun. Next is Saturn, 72,000 miles in diameter, the second largest planet, which is 886 million miles from the sun. Now we take another big jump to Uranus, 32,000 miles in diameter and is located the staggering distance of one billion, eight hundred million miles from the sun. Another big jump and we are at Neptune, 33,000 miles in diameter, located two billion, eight hundred million miles from the sun. And finally we come to little Pluto at the outer limit of our solar system, 3,600 miles in diameter and located three billion, seven hundred million miles from our sun.

Now we have had a fast trip around the solar system, haven't we? It has all been in our minds. We have needed no tickets, required no airplane or rocket, just our attention. How did we make this trip? Simply by the vibrations of our mind.

Now all this is not encumbered by our physical bodies. Our mind, and especially the "souls" of those who are trained, can move at will wherever it is wished. Does it seem quite so impossible, then, for us to contact other planets and the people on them? But more especially for "them," who are much more advanced than "we," to contact us?

Mr. Howard Menger will tell us more about our highly-developed neighbors, our older brothers of other planets and outer space: That these are truly our brothers, that they are not to be feared; that, instead, they are to be respected and treated with esteem and with as much understanding as we can muster. They are not to be treated with fear and hate, as we on Earth so ignominiously treat our own Earth brothers. People from outer space live in peace, not under the shadow of perpetual war.

But you did not come to hear me talk; you came to hear Mr. Howard Menger.

Howard Menger is a sign painter by trade, and a very good one, if growth of business is any criterion. I do want to stress this: Howard Menger makes no money from his contact with space people. On the contrary, it has cost him both a great deal of time and money. The fabulous amount of time he has devoted, and his great loss of sleep have naturally cut into his business and into his income.

Some people I know give one half of one per cent or even one percent of their income to charities, or to humanitarian undertakings, and feel rather exalted and self-righteous for having done so. But Howard's devotion to our space brothers has cost him more than 50 percent of his income, and, of course, his time. But in spite of these enormous sacrifices, Howard does not expect nor want anything for himself. He is making his sacrifice for his brothers in space and on this earth.

If one is prone to criticize, he might well remember this: Howard Menger is a most sincere person and a very fine gentleman. He has graciously consented to hold this meeting tonight so that those people who have seen him on the Steve Allen and other TV shows, and heard him on radio programs, especially the Long John Show which is on WOR (710 on your dial and on every morning from 1:00 to 5:30 A.M.), can see and hear him in person.

In the maelstrom of today the raising of men's minds must be increased, and truth brought forward. And it is Howard Menger's objective to do just this.

So without further ado, it is with great pleasure that I now turn this stage over to Mr. Howard Menger. . . .

COUNT-DOWN

Book One

Takeoff!

Book Two

Landing

"FASTEN SEAT BELTS!"

(Publisher's Note)

Before Howard Menger pushes the blast-off button and sets us spinning into uncharted regions almost beyond the stars, just a word, and a warning, from the Ground Crew.

This is Gray Barker, president of Saucerian Books, and the publisher of FROM OUTER SPACE TO YOU, giving you final instructions and wishing you a *bon voyage.*

I wish I were making this trip with you and living again for the first time your remarkable trip, as I did when I first read the author's manuscript; but alas, I must remain, as the publisher, with my feet on the ground.

To make this trip easier for you we have organized this book in a manner of which I do hope you will approve. Knowing that we will have casual readers, along with those who will wish to delve deeply into the more thoughtful metaphysics Howard Menger also will present, we have decided to publish the material in two books or sections. The first will contain the author's astounding story of his contacts and dramatic physical experiences. Not through intent, but by fortunate virtue of the fact that Howard Menger has an extraordinary command of narrative, you will find it reads as easily as any modern fast-moving adventure story.

We have moved the author's more thoughtful material to the last part of the book. The hasty reader may wish to finish Book One and put the volume on the shelf; but those who, after wetting their feet in the Milky Way, wish to gain deeper insight into the teachings of the space people as expressed by the author, will find Book Two even more rewarding.

And now, as you fasten your seat belts, our word of warning: this trip is dangerous. You will travel into areas of great intellectual peril, especially if you have made up your mind beforehand not to believe the author's story. Many of you will accept it as it is told; some of you may escape total belief and preserve many of your former opinions by believing that Howard Menger has presented only an allegory, as a framework for the metaphysical principles he wishes you to understand. Unfortunately there will be casualties, and for those we express deep sympathy. Some of you will read, and somehow, in an effort to disbelieve, will not give the author the benefit of even your good-humored laugh; instead the book will engender in you great anger and a desire to shout your disbelief to the world. Those readers we profoundly pity. For there are ideas here which will add a little something—a little inspiration, a few cogent thoughts, or even only a few thoughtful chuckles—to the life of the firmest skeptic.

I came back from the trip still a skeptic, and you may, too; but when you again touch Earth I believe there will be, as was on mine, a trace of a glow on your face.

And to those wonderful souls whom God, The Universe, or whoever He is, has blest above all: those who can believe without reservation, I give you a trip the likes of which you have never before dreamed possible, and a magic steed that will put a Pegasus to shame. You will enjoy the trip—ah how you will!

Finally, we add. . . .

What?

Oops! We've talked too long. It's too late to get out now! Fasten seat belts and get ready for——

BOOK ONE

TAKEOFF!

CHAPTER 1

THE GIRL ON THE ROCK

It was in Grantwood, New Jersey, that I met my first love.

I was born in Brooklyn, N.Y., Feb. 17, 1922, but when I was a year old Mother and Dad moved to Grantwood where we lived for several years.

The object of my affection was a classmate, a fluffy, feminine, blonde, blue-eyed bit of pulchritude, who completely stole my young heart.

But my parents were quite amused when I announced that I intended to marry my six-year-old sweetheart. The romance was short-lived and I brokenhearted when we moved to a country property in High Bridge, N.J. My brother, Alton, was four at the time; and I was eight.

I could not quite forgive my parents for taking me away from my love—until I saw the beautiful rolling hills and fields of northern New Jersey. The newness and excitement of the country slowly supplanted the hurt memories of the girl I left behind.

We children were delighted with our new home in the country. The fields were covered with daisies when we first saw the farm. To the rear of the house were fruit trees and woods and brooks which made a veritable paradise for two young energetic boys.

Dad and Mother worked hard to fix up the small, modest bungalow. Meanwhile my brother and I enjoyed long treks into the woods and fields. We played explorers, and exciting were our safaris into the jungles of the nearby woods, all fraught with imaginary and delightful dangers.

During the summer we had many playmates. Next door was a summer boarding house which accommodated many families during the summer season, and there we always found willing companions among the visiting children.

The winter I remember most as fields of white and steely blue shadows at about dark when reluctantly we would have to go inside, leaving behind our sleds, our games of "Fox and Goose" in the snow, and, rarely, skating when the nearby pond would freeze over. Somehow I remember those winters on the farm with a sadness and a gnawing, lump-throated longing, akin to the feeling when, as a child, I would, in quiet moments, even then begin thinking—that all of these wonderful moments could not last. That soon—far too soon, I would be a man; and even then I knew that once full grown, such moments and such mysteriously ecstatic feelings could never be re-experienced.

But I suppose we loved the summer best, when the woods and country living opened a whole new world to us.

It was in this pastoral setting, in the warm, lavish extravagance of June and July that I began to experience other feelings I was at a loss to explain.

I began to have "flash-backs," or hazy remembrances of scenes, places and happenings which somehow were familiar to me, but were outside my real experiences. They seemed to be of another world.

About this time we began to see the discs in the sky.

We watched them skim across the heavens, hover, and sometimes disappear. My playmates did not always see them, but I seemed to sense just when to look up. Alton saw them too, and when we told our parents they only smiled patronizingly as if in quiet agreement of what they felt were youthful flights of fancy.

Dad, a handsome blond-headed man with patrician features, was an adamant Catholic, while mother, with enough red in her auburn hair to insure a fiery disapproval of certain of his beliefs, was an unyielding Methodist; and sometimes the difference in interpretation of religious concepts would lead to most voluble discussions.

While their discussions never led to disharmony, I was caught between my loyalty to both of them and was often deeply disturbed because I did not know which to believe entirely. Religion was the only matter about which Dad and Mom could never agree, and I remember that when their talking of it led to near arguments I would go off to myself and sometimes cry.

But in confusion of trying to analyze, in my young way, these conflicting beliefs, I learned to think for myself and form my own concepts of God. That I believe I owe to my mother, for in spite of the differences of religious views to which I was exposed, she managed to firmly establish in my mind and heart the omnipresence of an Infinite Creator.

My brother and I continued seeing the bright, shining circular objects in the sky, and one day one of them landed in the field where we were playing.

It was a disc-shaped object about ten feet in diameter. Afraid, but fascinated and curious, we walked toward it to get a better view. As we neared it we noticed another bright object, a much larger one of similar design, hovering in the sky above the smaller craft, as if observing it and us.

Our hearts palpitated, but curiosity overwhelmed our fright as we proceeded cautiously.

When we were about 25 feet from the object, the larger airborne craft disappeared; and while we were trying to muster enough courage to go closer, the disc on the ground began vibrating, then took off at a terrific rate of speed in a blinding flash of light.

This experience we again enthusiastically recounted, but again our story was assigned to the realm of childish imagination.

I believe that Mother sensed, however, my gift of sensitive perception, for now and then when I mentioned such things to her I could tell that behind her pretended disbelief was a knowing look of understanding.

Gradually feeling a need to be alone, I began going off by myself, deep into the woods. There I had the feeling that somehow I could find the answer to the odd half-memories I experienced.

Some very strong impulse always drew me into one certain area of the woods. True, it was a beautiful section of the forest, idyllic in the summer, with a brook and almost tropical plants and foliage—but I knew in my heart that something else beside the natural beauty took me there.

A feeling of peace permeated the place, though interrupted at times by chattering and busy squirrels, running up and down trees in noisy pursuits. I remember timid and inquisitive rabbits, a deer which occasionally would come up very close. The soft-eyed gentle animal seemed to trust me and consider me one of the usual inhabitants of the forest.

To this enchanted spot I came at every opportunity I could find.

But one day in 1932, when I was ten, I saw something even more beautiful than the surroundings.

There, sitting on a rock by the brook, was the most exquisite woman my young eyes had ever beheld!

The warm sunlight caught the highlights of her long golden hair as it cascaded around her face and shoulders. The curves of her lovely body were delicately contoured—revealed through the translucent material of clothing which reminded me of the habit of skiers.

I halted in my tracks, and for a moment my breath stopped. I was not frightened, but an overwhelming wonderment froze me to the spot.

She turned her head in my direction.

Even though very young, the feeling I received was unmistakable.

It was a tremendous surge of warmth, love and physical attraction which emanated from her to me.

Suddenly all my anxiety was gone and I approached her as one would an old friend or loved one.

She seemed to radiate and glow as she sat on the rock, and I wondered if it were due to the unusual quality of the material she wore, which had a shimmering, shiny texture not unlike but far surpassing the sheen of nylon. The clothing had no buttons, fasteners or seams I could discern. She wore no makeup, which would have been unnecessary to the fragile transparency of her camellia-like skin with pinkish undertones.

Her eyes, opalescent discs of gold, turned their smiling affection on me with a tranquil luminescence.

"Howard," she spoke my name, and I trembled with joy.

"I have come a long way," and she paused smilingly, "to see you, Howard . . . and to talk with you."

I shall always remember those first words exactly as she spoke them; but then my thoughts swirled in a maelstrom of emotion and slowly coalescing understanding as she continued to talk.

I remember that nobody had ever spoken to me as she did. She talked with me as if I were much older.

She said she knew where I had come from and what my purpose would be here on Earth. She and her people had observed me for a long time and in ways I would not quickly understand.

When she spoke of her "people" I still could not understand they were from another planet; as I listened in awe, my eyes delighted in feasting on the beauty of this lovely creature.

Every movement of her body, as she stood up and walked toward me and reached out her hands to me, was a symphony of rhythm, grace and beauty. I seemed to be encompassed by the very glow, almost visible, that emanated from her presence. Somehow the entire area surrounding us appeared to take on a greater kind of radiance. I have often tried to describe it as like seeing a Technicolor movie in three dimensions and being a part of the action in the film.

Again she pronounced my name and reassured me she knew who I was, "from a long, long time."

And then some words that have taken on even more joy and meaning as I have grown older:

"*We are contacting our own.*"

She told me that even though I did not understand many of the things she told me then, later in life I would. Her words would be impressed on my mind—I suppose she said "subconscious"—but it was difficult, as she said again, to make me understand.

I remember that she compared the idea to that of a phonograph, which would be played back to me time and again.

"It is no fault of yours, Howard, that you cannot understand everything. Do not worry." And she laughed musically.

She continued to speak to me as if I were an adult. I cannot remember many of her exact phrasings, but the "phonograph" has played back the ideas, each "replay" taking on more and more meaning. Some of the actual words were beyond me, for they were words that meant nothing to a ten-year-old: "frequency" . . . "vibration" . . . "evolvement." . . .

She smiled most of the time as she spoke, and now and then she laughed as she answered questions before I could ask them. She seemed to know all of my thoughts.

But then a look of sadness came over the beautiful face, and tears came to my eyes as for the first time I pitied my new wonderful friend.

She spoke of a great change to take place in this country as well as the world. Wasteful wars, torture and destruction would be brought on by the misunderstandings of people.

"As you grow older," she said, "you will grow to know your purpose. You will help other people grow to know their purpose too."

This would depend on "evolvement" and "universal laws," and I would be drawn to other people who have missions akin to mine.

Then she stood up and I knew she was about to say goodbye. I noticed she was about my mother's height, slender, lithe, with no exaggeration of voluptuous curves.

She extended her hand and grasped mine. It was warm and soft and I was reluctant to let it go.

I began to cry.

"Don't worry, Howard," she promised "You may see me again . . . but it will be many years before you do. And I am not nearly so wise nor wonderful as others of my people who will often visit with you."

"Where do they live," I asked perplexedly and almost petulantly.

"Ah, far away, but you will find them. They will come to you. You will know where to go and meet them. And if your mind is troubled, remember, they will always be around—watching out for you . . . guiding you."

Again she laughed and I could not help be affected with her happy humor. I laughed, too, though with tears drying on my face. She said I should leave first, then she would go.

"May I look back?"

"Oh yes, Howard, you may look back!"

And I did, after walking slowly away. She was still sitting on the rock, smiling, and she waved.

I turned and ran, sobbing, first hardly audibly, then louder and louder, till my wails of a happy kind of sadness grew and filled the forest.

CHAPTER 2

THE MAN IN KHAKI

I often went back to the brook in the woods, hoping to see her.

The place looked the same, though lacking the radiance which seemed to illuminate it that one day. The brook still ran musically beside the rock; the foliage was still lush, and the squirrels kept up their chatter—but the lady was not there.

But in time it seemed that the enchantment of the place had gradually faded. Perhaps I was growing up. Perhaps I had never seen her and only imagined I did.

I wondered about it often as I lay awake nights, remembering. I decided that even if the beautiful girl were not real, the things she had told me were taking on more and more reality.

I remembered she said I would undergo many trials. That I would be unhappy. Partly due to the many mistakes I would make—natural things, due to miseducation; or the pangs incurred in a gaining of education.

The experience with the golden haired naiad of the forest had a profound and lasting effect on my life.

Throughout my life the things I had learned in the forest were to lead to conflict with the conventional ideas of the world.

It began with a difficult time in school. A great deal of the information my teachers tried to convey to me was, I knew, untrue.

The girl on the rock had told me of life and people on the other planets; yet in school we were taught that the planets in our solar system were lifeless worlds, either too hot or too cold, or covered by poisonous gases. Nevertheless I soon learned it was often better to put down the accepted answers even though I knew they were wrong. One had to live with other people and their ideas.

But sometimes I rebelled, and as a result many of my classmates and teachers grew to think I was odd.

My intellectual rebellion had a bad effect on my school grades.

I remember that once we were assigned to write a theme for an English class, and I chose the subject, "The Evolution and Evolvement of Man," in which I developed how man had ascended from the very life cell (which in itself possesses a consciousness and a portion of the Infinite Creator), to the time when he first stood erect on two legs. My teacher marked an "F" on my paper because my views conflicted with his religious ideas. As in other classes, I noticed I was becoming unpopular.

So I became quiet and retiring, keeping my ideas to myself.

I finished four years of high school despite my frustration of being unable to express what I knew to be true.

Shortly after leaving high school in 1941 I worked at an arsenal in north Jersey for more than a year. Then I entered the army.

This was 1942.

They sent me to a tank outfit in the southwest.

For a while the harsh new environs of Army life took my mind from the many things which had occupied it while growing up. We were on maneuvers from Texas to Louisiana for 18 months. After arriving there, somehow the girl on the rock and the ideas she had expressed, particularly of the brotherhood of man, grew to be like a dream, as the terrible realities of war pushed her from my thoughts.

But, then, in one of the most desolate places I have ever been, a firing range outside El Paso, another strange thing happened that once again brought my earlier experiences into vivid focus.

We were camped in the desert, not far from the Rio Grande River. The naked, stark sterility of the immense scene nevertheless had a kind of rather odd beauty about it. It was a lonely and silent place. We could hear coyotes barking in the distant hills, which, though in contrast to the immense tranquility of the region, they seemed only to add to the great silence by providing a contrast to it.

During many of those nights I had a feeling I did not dare describe to my buddies—a feeling that we were not alone. That we were being observed . . . though by watchers who protected us.

One night I again saw what I now know to be an observation of discs in the sky, and thereafter I saw more of them, both during the days and nights.

The results of past experience restrained my exhuberance, and impatience to point out the discs to my buddies.

Whether or not they saw them I do not know; if they did, no doubt they mistook them for our own high flying aircraft.

One night a couple of my buddies insisted that I go with them into the nearby town of Juarez. Though I did not appreciate the loud, garish entertainment offered by such towns, I agreed, hoping it would break the monotony of camp life.

When we got into town I had no difficulty breaking away from my friends, for they knew I was not interested in the kind of entertainment it offered us. I wandered off by myself to look for some souvenirs to send home.

As I walked down a street toward a curio shop I had spotted, a taxi pulled over to the curb and the driver addressed me in Spanish.

I replied in perhaps the worst Spanish on earth that I did not speak the language and his look showed me he hastily agreed. Then he said something else and pointed at a man in the back seat.

I am afraid the novelty of the occasion quite nonplused me. Much to my later chagrin I remember the first thing that struck my mind were some of the stories related in the bull sessions back at camp.

The man had long blond hair which hung over his shoulders. His skin appeared suntanned. The first quick observation showed that he was taller and heavier than the average Mexican.

He spoke to me in English quite pleasantly, though I remember he had a slight Mexican accent.

"I have something to tell you. Would you get in the cab?" he asked, but I demurred, making the excuse I had to find my buddies and go back to camp; and at the same time I walked on. As I turned he smiled and merely said, "All right," again quite pleasantly.

When I told the others about it, there were many guffaws; and for several days I was the butt of much good-natured ribbing.

But upon reflection I wondered if I could have made a mistake. I remembered the discs, and again many of the things the girl had told me. Could this have been some of "our people" who she had promised would seek me out:

". . . They will always be around . . . watching out for you . . . guiding you."

But I never saw the man again.

Maneuvers moved us from place to place and finally we

arrived at Camp Cook, in California, where a lot of scuttlebut had it we were being readied for shipping out. I wangled a leave of absence and went to Abilene, Texas, to see my first born—my son, Robert. My wife and baby were staying there with relatives, and we had a happy reunion.

A few days after I returned, another happening in the chain of events, which would finally change my life completely, occurred. As I was walking on the camp grounds I heard someone call my name.

I looked around, but saw no one familiar. I thought I must be mistaken.

As I continued walking, a man in khaki uniform approached me from the opposite direction, and I again heard my name called out.

It seemed to be coming from his direction, though I could not figure why because I did not know the man.

He was of average height, and apparently muscular and well built.

All the while I was puzzling over what he wanted with me and the peculiarity of the voice—not so much the voice as of my confusion about where it was coming from. It seemed to be coming from his direction, yet it was not audible. Later I was to learn the sound was not audible but a projected mental sound which I only thought I heard.

A stab of memory quickened my mental processes as I realized this was telepathic communication, for I had heard the sound not with my ears but with my mind.

I stopped in my tracks. Was this the kind of man the girl had spoken of—someone from another planet? This sudden realization that such a thing was possible stunned me for a moment and for a brief second or two I was even afraid—though it was an occasion I had long hoped for and longingly expected.

Then he greeted me, speaking audibly, pronouncing my name and extending his hand. I stood there staring at him with, I am embarrasingly afraid, a very blank look on my face. Slowly I raised my hand and took his.

I suspect the first space man I met did not gain an impression that earth people had firm handclasps, because mine was very weak.

Then he smiled, put a gentle pressure on my hand, and I suddenly felt warmth glowingly permeate my entire body.

Then I returned the handshake, grasping the handclasp with my other hand as I again realized some of the same feelings I had experienced long ago on the rock in the woods.

As that scene came flashing across my mind again in what seemed a whirl of stimuli, he picked up my thoughts.

"Yes, Howard, I know of the contact you had with one of our people when you were very young, and you will see her again in the future. . . ."

I looked up with what must have been an obvious appearance of joy and met his eyes. He smiled knowingly.

He was a fine looking man. Although there was something definitely unusual about him, he could have passed—and did—for an ordinary G.I. The singularity of the man probably was not because of the finely chiseled features and the luminous, almost liquid quality of his eyes, but in the communication I felt. I could sense that the man was kind, wise, emotionally and spiritually developed beyond anyone I had ever met.

Although a kind of reserve he wore as if a part of him set him apart from an ordinary person, I somehow accepted with no surprise the emergence of an underplayed, yet natural sense of humor.

"I know about your Juarez contact," he said; and it was confirmed that the man in the taxi had been one of the space people.

He chuckled.

"We told him he should cut his hair. I have, you know. It's difficult even for us to keep up with you folk and learn just how you think."

I apoligized for goofing up the contact, but he waved away my remonstration. He realized that army regulations encouraged caution in such areas, and that Juarez was not the best place in the world for an interplanetary meeting.

Then I stood open-mouthed in amazement as he related, in a matter-of-fact manner, things I had never dreamed could happen.

Many Mexican people knew about what I termed "flying saucers" and had been contacting the occupants of the craft.

"Long before the time of the Conquistadores," he added, "we made contact with the Aztecs. We helped these people in many ways, and it is too bad the conquerors came in war instead of good will and friendship; for there were many things the Aztecs could have taught them. Instead they witheld these secrets, and these perished with the civilization."

Some of the secrets had to do with the use of sound and light to produce power and run machinery, though my new friend didn't elucidate. He remarked that gold discs which were sent back to the Queen of Spain contained such secrets, but the Spaniards were interested only in melting down the gold. I gathered from his conversation that the discs were some sort of sonic instruments used for levitations when tuned to the frequencies of individuals using them.

Other civilizations received the use of marvelous instruments, and these were used for peaceful purposes. But as in the case of the Aztecs, the secrets were destroyed or forgotten when warlike races invaded.

"Thus it happens, over and over again, Howard. You'd think we'd give up—we won't."

The man in Juarez was a visitor from a planet (he did not say what planet) who came to contact remnants of his own people still living on earth—descendants of an ancient race which originally came here from his own planet.

The surprises kept coming. He suddenly told me that my outfit would be leaving for Hawaii soon, and that I would be put on detached service with special duties which would give me more free time for certain tasks I was to perform. He said I would have a contact in Hawaii and would be given further instructions.

Another person in our camp had also been contacted, he said. I asked him who.

"An Army officer," he replied, without giving the name. Sensing my curiosity, he added. "It makes no matter; you and he will not meet."

A few weeks later we shipped out to Hawaii.

CHAPTER 3

HAWAII CONTACT

The "G.I.'s" predictions proved remarkably accurate.

After being sent to Hawaii, as he had promised, I was taken out of the tank crew and transferred to Battalion Headquarters and made a battalion draftsman; and, as predicted, I did work on detached service with the Navy.

As we had parted I could not help thinking that these people from other planets seemed to know the past, present and future. Again he had sensed my thoughts and smiled, terminating the conversation with another handshake, and walked away.

Most everything he had told me had already come true, except the contact he promised, and this I impatiently awaited, almost breathlessly.

It was a strange, wonderful feeling to meet these people. Somehow, as unimportant and weak as I felt in their presence, there was still the knowledge of kinship I couldn't help sensing.

So it was that one early evening after work I did not hesitate to accede to a strong impulse to visit a section of cavern area a few miles away.

I "borrowed" a jeep and took off.

I didn't know exactly where I was going, excepting for the general area. It seemed I was being led.

Near the caverns I stopped, then pulled the jeep off the bumpy, dirt road, and walked through the dense underbrush toward the caves.

I knew I would meet one of the space people. Ordinarily I would have been fearful of being alone in such a wild place. But the thought of the meeting erased all of my natural apprehensions.

Suddenly I halted as I saw a figure ahead of me. Through the underbrush I could see it was a female form.

As I walked closer I discovered she was a beautiful woman with long dark hair and dark eyes.

She was dressed in a sort of flowing outfit of pastel shades. Under a kind of flowing tunic, translucent and pinkish, she wore loosely fitted pajama-type pantaloons.

She stood about 5'6", with the dark, wavy hair falling over her shoulders and the tunic floating gracefully around the shapely contour of her body. The warm, moist air of the tropical evening seemed to caress her finely molded features.

I stopped in my tracks, staring at her in uncontrolled admiration, until she extended her hand and called out my name.

Although I shall always remember the girl on the rock with a special kind of memory, this girl, too, exuded the same expression of spiritual love and deep understanding. Standing in her presence I was filled with awe and humility, but not without a strong physical attraction one finds impossible to allay when in the presence of these women.

She immediately sensed that part of my feelings and also my embrarassment at knowing that she sensed them.

"Oh, Howard," she almost chided, "it's only a natural thing, I feel it myself. It flows from you to me as from me to you."

But many other men under similar circumstances would not react in the same gentlemanly manner as I did, she remarked, as I could detect an undercurrent of good-humoured jest in her words.

Then she grew more serious.

"That is one of the reasons you were chosen out of many thousands for contact with my people and the enlightenment you will consequently receive."

Again she read my thoughts.

"Oh, to be sure, Howard, if you *weren't* a gentleman, I would have the proper defense. So many people's egos are greater than their humility. But yours isn't."

Again I was taken aback with amazement at the knowledge of these space people.

"I know about the little Portuguese girl and what you did. It was a wonderful thing to do, Howard, and it showed you as the real man that you are."

I am always sensitive to praise, though I deeply enjoy it. I shyly lowered my head and turned a bit red as usual. She referred to the little blond girl on which some of my aggressive buddies tried to force their unruly attentions. I had stepped in, suddenly brave enough to fight a mountain lion, managed to extricate the young woman and had taken her home. Her family had greatly appreciated the gesture and had received me into their family as an intimate friend.

"In other words, I think you're 'passing,' Howard! Isn't that the way you say it in school?"

I was again overjoyed. I was so afraid that, feeling so powerless and inconsequential in the presence of these people, they would think me as inferior as I myself imagined.

"You have been observed closely, as you now realize. You will be trusted and have further contacts."

She also made predictions. Our outfit would go to Okinawa, would arrive there between April 1 and 5, 1945.

My abhorrence of war she easily picked up telepathically.

"I know how you feel, and it is most admirable. You cannot think of killing a living soul. But yet you cannot understand why you yet help play such a role. You will be contacted on Okinawa, and you will be told much more about this."

I hesitated to ask her if I might be killed, but it was on my mind.

"Oh no, don't worry—but be careful! You will have some narrow escapes."

The average person with whom I talk about these contacts does not realize that the space people, though far superior to us physically, mentally and in spiritual developments, are still much like us. Often little gasps of amazement come when I tell of intimate conversations, and the warm humor of the visitors. They would probably stand sanctimoniously before the space people, afraid they might say or think something wrong—until, of course, they received the same feeling of ease I did immediately, even at the first meeting.

At such a meeting one knows innately that one's every thought is bared under powerful telepathic observation. And with such knowledge one suddenly realizes he cannot hide anything, and becomes completely honest, both with himself and the visitors. It is a refreshing, cleansing feeling, which carries over into everyday dealings with one's fellow men.

The conversation with the beautiful girl was so fascinating I hoped I hadn't annoyed her with too many questions. I learned, for one thing, she was from Mars. As to meeting her again, she wouldn't state firmly; instead she explained we might meet again, and I would have to know by my inner feelings whether it was really she.

Suddenly I realized the sun had set, and as I looked toward the horizon, still bright with a hundred shades of red, then back at her, she smiled, and extended her hand.

We said goodbye and I walked back to the jeep. It was dark by the time I arrived back at camp.

CHAPTER 4

A NARROW ESCAPE

True to the girl's prediction, we landed on Okinawa the first week of April, 1945, and into a reign of horror she had charitably spared my anticipating.

It is an indescribable feeling to board an LST and head for an enemy shore. As our small boat neared the beach, I steeled every nerve in my body, not knowing when the fury of enemy resistance would be unleashed.

At that climactic moment the briefing we had just received on board ship held little comfort—but we hoped they were right! Our aircraft had given the island a saturation bombing of such thoroughness and intensity our officers believed that all organized surface resistance had been smashed. And even as we neared the beach our naval vessels riddled the island with concentrated shell fire.

I had the feeling our landing had been too quiet. And I was right. The Japanese were still there—in force. They were really "dug in," hiding in caverns and concealing themselves in outlying areas.

If I had ever thought of war with a connecting glamour, that idea was soon gone. War on Okinawa was a grim, horrible thing, without a vestige of glamour, parading and glow some people may associate with it. This was guerilla warfare, without any actual front, with hand-to-hand fighting.

The Japanese had not lost their power to retaliate. One evening after a rather quiet day, our bombing and strafing planes were returning to the airstrip when a Navy Hellcat, the last plane in, flew in low, without attempting to land. Suddenly it opened fire on us. We hit the ground and took whatever cover we could find as it sprayed the area with machine-gun fire.

We were surprised and aghast. Had one of our own pilots gone mad? We later discovered a Japanese pilot had somehow gained control of one of our planes and had managed to slip in on us and make the daring attack, which did great damage.

The enemy kept us constantly harassed by shelling us from a nearby island which they still held. The shelling did little real damage and was done I suspected, mainly as a psychological weapon.

One day as I was patrolling near the air strip one of the shells fell short and hit pretty close to me. I heard it coming and flattened, and thought I had escaped injury.

As I got to my feet I felt something stinging in my right eye. I put my hand to my eye and managed to pick out somethings with my fingers. It was a piece of shrapnel.

I stumbled to the hospital area and received treatment, but the eye became infected and finally went blind.

I was hospitalized in a large tent near the camp where busy doctors and nurses worked hard and skillfully at all hours; but they could not prevent the infection from spreading to my other eye. I was completely blind.

Something happened in the hospital tent that I have often wondered about. Perhaps I can never be certain.

During the first week there a kind, soft-spoken woman came to my bed, and began talking with me.

When I asked her if she were a nurse, she didn't reply directly, but said she was not really assigned to my section.

"You are one of the persons I have come to see."

I detected a reluctance in her voice to tell me much about herself, and did not press any more questions.

She must have known a lot about me. She called me by name, though I figured she could have obtained that easily from the hospital records or the doctors in charge. She offered to write letters home for me, which I declined, not wishing my family to know I was hospitalized.*

She assured me my sight would be restored, and, surely enough, it came back gradually. When I first saw my soft-spoken friend, I noted an attractive woman with wavy brown hair, dark eyes, and fine white teeth. She was dressed in an army nurse's uniform.

Although I suspected she was one of the space people, she never made herself known directly. Near the time of my release she said that I would soon meet a very interesting person. I assumed it would be another contact.

I never saw her again after that day on which she said that.

When I returned to camp my buddies suggested that I apply for a purple heart and that by the point system could return home more quickly. I laughed and replied I certainly did not want any medals, and did not want to go home yet. While my buddies kidded me about not wanting to leave the pretty nurses I had met at the hospital, I smiled, knowing

* When I returned home I found that my wife had known almost the exact date I lost my sight. She had told her family, "I know what happened to Howard. He's blind!"

that I indeed had some unfinished business, but of a type they could never conceive or dream of.

If my hunch I had about my pretty friend at the hospital was true, there was someone else on the island I had to meet.

. . .

Two weeks after my release from the hospital I received an impulse to drive away from the camp, never anticipating what would really occur. Again I "borrowed" a jeep, my heart leaping in anticipation of another meeting I was certain would result.

I drove along a dirt road toward the northern end of the island, passing deserted native villages which had been bombed and shelled almost out of existence. The road let down into a valley where the trees and shrubs were still intact, indicating there had not been much shelling.

I drove off the road through the light shrubbery. Toward the hills I saw evidence of a number of caves. Closer inspection showed some of them had been dynamited to seal in unfortunate Japanese who had "holed up" there.

About 500 feet off the road I stopped and for the first time was perplexed.

The impulse had left me.

It had been strong at the beginning, but now I was left without further mental instructions.

At the same moment I was struck with the realization that it was growing dark.

This time, as before, the impulse to drive away from the camp was attended with an unusual kind of bravery. But now I realized the spot I was in.

I would probably lose my way if I tried going back. Then the remaining Japanese on the island were bolder after dark. My jeep might be attacked or a sniper might get me.

I decided the best thing to do was to spend the night there.

I unrolled my pack and pitched a tent. Finding a piece of wire nearby, I strung it in a square ten feet away from the tent and about one foot above the ground. One end I attached over the tent, with my mess kit poised directly over my face, so that the slightest pressure on the wire would dislodge the contents of the kit onto my face and awaken me. After setting my crude but effective alarm system, I slipped off into sleep, while instructing myself that I must arise early enough to make my way back to camp before I was missed.

I had not slept long when, with a clang, the mess kit came down in my face. I instinctively grabbed my carbine.

Cautiously I peered out of my tent. It was a misty night, with a few overhanging clouds. I could barely discern a shadowy form moving around outside. Then my eyes became accustomed to the dark and I could see it was a large animal of some kind. I was greatly relieved. I figured it was likely one of the stray domestic animals from one of the deserted villages.

So I reset my alarm and once again fell into a deep sleep.

I must have slept for two hours when, BANG! The mess kit again was in my face.

Just as I reached for the carbine, a bayonet came down through the tent between my left arm and the gun.

In a split second I had full cognizance of my precarious situation. I bolted out of the tent, and there over it was the hugest Japanese I had ever seen. He was so busy harpooning the tent (and what he hoped was me) that he did not see me emerge.

I let him have it in the back of his head with the butt of the carbine, and he crumbled over the tent.

Just then two others came at me with bayonets poised from a few feet away. Never before had I realized my strength and agility. I lurched toward them, grabbed the rifle from the one on my left, and it went off into the air. At almost the same moment I shoved the butt of my carbine into the face of the other one.

It all happened so quickly that I actually collided into the one on the right and fell. As I stumbled, I rolled away from him, picked up his rifle and gave him a crashing blow on the head.

He slumped to the ground with a groan. The other man took a shot at me, but missed me in the dark. I took a running dive at him. I was now unarmed. With a quick, unexpected butt in the belly with my left knee, I managed to get his rifle away from him and then hit him over the head with it. Down he went in a heap like a sack of laundry.

I thought I had killed all three of them, yet during the heat of the skirmish I was overpowered with the feeling that I must not shoot or stab them. I sat down for a moment, breathless at what had happened. I think I cried unashamedly because I feared I had killed three human beings. After I calmed down, I rolled the first Japanese off the tent, got my gear together and left.

I found my way back to camp without incident, and not until then did I realize what a narrow escape I had. It was fortunate that the three Japanese were very large; and I noticed that these larger men were not as quick on their feet as the smaller soldiers. As I was about to credit myself with the remainder of the victory, I suddenly thought of the meetings I had experienced. Then I realized that I probably had a great deal of help, and very capable help indeed!

CHAPTER 5

A PREDICTION

Back at camp the very next night I awoke with a start.

A voice called to me. I thought it must be one of my buddies. The voice called out, "Howard," several times.

I looked around. Everyone was asleep.

The voice came again.

"Howard!"

Then I realized it was the same inaudible kind of communication I had received at my second contact.

Again I received a strong impulse to drive to the northern end of the island, and I dressed quickly and silently left the camp.

The impulse again led me to the location I had been the night before. I got out of the jeep and walked toward one of the caves.

Then I saw him, standing near the cavern entrance.

I could make out in the moonlight that he was very tall and well built. I walked toward him. He must have purposefully stood in the light so that I could see that he was Caucasian, dressed in Army khaki and unarmed. Even though the events of the previous night still cautioned me, I had no doubt from the first sight of him. I had the same warm, comforting feeling I had experienced at the previous meetings.

When I was within speaking distance he smiled and said, "Hello, Howard."

I returned the greeting.

"I see you got my message and followed the mental directions all right this time. It's too bad about last night."

"Then you know what happened?" I asked unbelievably.

He nodded.

"I'd rather not talk about it," I told him. The horror of having killed three human beings whose bodies were probably nearby, made me feel uneasy and disturbed.

"Yes, of course I know—and understand," he assured me. "In fact I know more things that you have done and will do than you realize. We know more about your people on Earth than your people know about themselves. That's how I can understand why you don't want to talk about the incident."

He sat on a rock and indicated that I sit by him.

"You see, my brother," he continued, "this experience was necessary, as you shall later discover. If you had known of these contacts and the message we bring sooner, you would not have entered the army with the purpose of killing your fellow man.

"Knowing of your reluctance to kill a human being, we thought it best for you to go into the army as a regular G.I., because you would thus be a better contact for us."

Then I began understanding why it had been necessary for me to go to war.

"If you had been aware of the complete futility of killing others in wars you would not have gone into the army and probably would have been imprisoned as a conscientious objector."

I was beginning to see that these people were as realistically practical as they were highly developed spiritually.

"Such a record would not have been good background for the job you are yet to do. People would look upon this background as unpatriotic and would not listen to you."

My friend then told me that contacts are chosen on the basis of what the individual is like deep down inside: what he will do under duress and in extreme emergency; whether he would rather kill or be killed.

"Actually, Howard, there is no death. Only the physical body, or the shell, dies, and even that is not really destroyed. The soul lives on eternally, learning by its mistakes, always progressing. The good that is done is accredited to that soul. The mistakes are forgotten."

I asked him his name so that I could address him.

His answer was simple:

"Names are not very important."

The chain of events had been so unusual that I had often wondered if I were only having a dream. I asked him if what I was experiencing was really true.

"Am I losing my mind—or is it really possible that I am contacting people from other planets?

For the first time my friend was amused—but sympathetic.

"Yes, you have been contacted by people from other worlds and you are not losing your mind."

Then with an almost confidential, but humorous tone, he added:

"If you think you're crazy NOW, Howard, wait until you

see some of the other things that are going to happen to you!"

He told me my country would win the war.

"Without the aid of the United States, Britain and Russia would have been conquered by the Germans, who are far ahead in technical and scientific skills, and—(he paused)—you are yet to learn many things about their developments.

"They have used this knowledge for destruction rather than for peace in your world. The Japanese will surrender shortly, for they are about to be blasted into submission by a power which will shock the world, both in amazement and in sensibility. It will be a more infamous kind of destruction than occurred at Pearl Harbor—far, far more infamous."

This same power, he explained, would be used partly for peaceful purposes by governments of the world, but mainly for defense. It was the latter use which he said could lead to the destruction of the entire planet.

"Remember that the murdering of one man or mass destruction is one and the same. Evil is not gauged by the magnitude of the crime. The evil intent which goes against our Infinite Father's laws will turn the same laws against your people. However, our Infinite Father does not punish, destroy, or put any discomforts in our way. Man himself creates such, and blames God for his own discomforts. He even asks the Creator to punish others for their wrong-doings."

He paused, and I was ready to ask a rather selfish question, considering the magnitude of the instruction I was receiving.

Like the others, he anticipated my question.

"Nevertheless, Howard, the war will be over soon and you will be home by Christmas."

He then told me other men on the island had been contacted, but that none of them knew the others and each kept the secret.

"Be calm and steady yourself, Howard. We have been spending a lot of time in conditioning you and preparing you for your work to come. We are contacting people all over the world."

He said that something very shocking would occur soon which would shake the world from its lethargy and raise it from the shadows of ignorance into the light of awareness and understanding. But this great event would not come without much misunderstanding, resentment and hatred.

Nevertheless, we can learn only through mistakes, he reiterated.

"Man must learn what he is, where he came from, and what his real purpose is here on this planet."

He told me I would be further enlightened as to my true mission, and as I listened I inwardly felt I had begun to know what that purpose was.

"Not everyone comes into the realization of this, but the ones who have been contacted are aware of the true purpose of mankind."

He moved, as if he were rising, then paused.

"I have almost forgotten, Howard, that you are likely very curious about many things, more about us, for instance. You're free to ask, you know. Certain things we cannot reveal to you at this time, but I'll do my best to answer your questions."

Relaxing a bit, I put to him some of the questions I was almost afraid to ask earlier.

He was from the planet Venus, he replied to my most eager question.

"But how do you come here?"

"In a ship. A ship unlike anything you have ever dreamed of. The force will be difficult and probably impossible for you to understand. It is an electromagnetic force, not unlike the force which holds planets, suns, and even entire galaxies in their orbits. This force is a natural law, which has been given to us by our Infinite Creator to be used for good purposes."

"But why have not our own scientists discovered this power."

"Ah but they have. Yet they do not know how to apply it. If they did know the secret they probably would use it for destructive purposes. Until they are ready to utilize it for peaceful ends, our Infinite Creator will prevent their understanding it."

We talked for what seemed to be but a short time. Later I realized it had been more than an hour.

Finally he arose.

"We must end this discussion, and I will say goodbye to you, my friend," he said softly as he extended his hand and smiled.

"Will I see you again?"

"No, I'm afraid I'm done with you—and I'm sorry, because I like you. But you will meet others who will continue this instruction. Your contacts will become more frequent back in the states. We have much work to do on your planet among your people, and we must do it quickly—*while there is still a planet and people to work with!*

He turned as if to leave, then halted and addressed me once again.

"You will wonder what I meant by that last remark about danger to your planet. Very shortly you will know to what I referred."

I left the area, somewhat confused and distrubed. I could not believe that conditions were so serious on our planet.

A few days later someone pressed a button and fiery hell fell on Hiroshima. . . .

CHAPTER 6

BACK TO THE STATES

We left Okinawa in late October and boarded a ship for Korea. From there, where I stayed a very short time, we shipped out for the States.

I arrived on the west coast about the middle of December, 1945, and wasted no time boarding a train east. My wife and I and our small son enjoyed a wonderful Christmas that year in the homes of our parents. After the holidays I found a house for my small family near Washington, N.J., and we moved there.

Like most other young men, I looked at civilian life almost fearfully. For the first time I faced the responsibility of providing for a family in a changed world.

I considered several jobs, but couldn't decide which to make my livelihood. In this period of readjustment and confusion which was fortunately short, I don't remember thinking to any degree about my space contacts. Early one morning an idea struck me.

"I'm going into business for myself," I remember telling my wife.

I possessed some skill as a sign painter. So I bought a second-hand truck and some equipment, rented a shop and soon was ready to do business. Although I had fretted with various misgivings about going into business, soon I found a great deal of work, and noted, with gratification, that my customers liked the services.

One day I realized I was doing very well, indeed. I had taken care of my debts incurred by going into business, and was putting a little money in the bank.

My life had settled down to a normal pace and we were content.

For the first time I found myself having time to think about the strange experiences I had encountered overseas, and again remembered that first wonderful meeting during childhood.

It was in June, 1946, when I again received a very strong impulse to return to the enchanted area of my boyhood. I drove to my parents' home in High Bridge, got out of the truck, and headed for the wooded area remembered so well.

As I walked, I began wondering about my experiences and caught myself doubting that the events actually had occurred. Somehow, the war, and the unfamiliar places I had been, now seemed unreal—as if it had been a movie I had attended the night before and then found myself back in the world of reality.

For a moment I even wished it were so. I had a good business, a family, and for the first time in my life was content. The memories of the girl on the rock and the other wonderful people were like fascinating dreams—but with them a feeling of responsibility, and a certain knowledge that to fulfill whatever destiny they had mapped out would entail an upsetting of this happy, easy life.

I paused and looked around me. These woods and fields were real—like they had always been. They were more real—that was the tragedy. The enchantment of my boyhood was no longer here. Now I lived in a world of men, a world of hum-drum average men. I found myself liking it; too much.

I kicked at a rock and watched a small beetle scurry from the depression where it had been. I watched it for a second, then turned. I had decided to go home.

Then I jumped a foot off the ground and let out an exclamation. There was a tremendous flash of light and the sense of heat on the back of my neck. I turned. Above the vast western section of the field a huge fireball moved at tremendous speed.

It looked like a huge spinning sun, shining, pulsating and changing colors. It hovered over the field, as I stood watching it, seemingly transfixed.

The pulsating color changes diminished and the fireball turned into a metallic-looking craft, surrounded by portholes.

It descended slowly to the ground. When it was almost on the ground I could make out the form clearly. It appeared to be bell-shaped, and reflected the sunlight like a mirror.

I realized this was not a machine built by anyone in this world.

I didn't know whether to run back to the truck and get my small camera, drop flat on the ground, or get out of there entirely. Suddenly encountering such a thing after my quiet life back at home was frightening.

Then I remembered my wonderful friends and knew that some of them must be inside the machine. So I waited and watched, still unable to move, completely fascinated.

Soon an opening appeared on a flange around the bottom of the craft. It is difficult to describe the opening, because it wasn't there one moment and the next moment it was. The best I can describe it is to compare it to the opening of an iris in a camera lens.

Two men stepped out.

They were dressed alike, in blue-gray ski-type uniforms. They wore no hats, as long blond hair moved with the breeze. I could see they were fair-skinned, and of average height. As I had admired others who had contacted me, I noted they were physically handsome, with broad shoulders, perfect proportions.

Then something removed my attention completely from the two men.

Through the large opening in the craft stepped a beautiful woman. She had long blond hair and was dressed in a similar outfit, which fitted loosely over what was a shapely body. The material was semi-translucent, of a soft pastel color which seemed to glow.

She looked at me and began walking toward me.

My heart began palpitating as a shock of remembrance stabbed at me.

"Could it be? No, is it possible!" I thought.

And yet, as she approached, she appeared to be the same woman I had met 14 years ago. Now a man, I could really begin to appreciate her beauty.

She smiled, and seemed to know what I was thinking.

But I couldn't put the amazement out of my mind. This lovely creature had not changed at all in appearance during those 14 years. She still looked only about 25.

She approached me, extending her hand. As I took it, a feeling of relaxation and well-being consumed me and, for the first time since seeing the fireball, I could move.

"Are you actually the girl—the girl on the rock?" I asked.

"Yes I am. The same girl, Howard."

"But you're no older—"

"Oh but I am. Guess, Howard, how old I really am."

I just stood there looking at her.

"I'm more than 500 years old. Now you can refute anyone who says a woman tells little falsehoods about her age!"

"But you haven't changed—"

"Of course not."

Then she looked at me, at my entire body, and my face burned. It was something like a visiting relative looking over a small boy to see how much he had grown.

I knew she was gently teasing me as she winked and added, "Oh, but YOU have changed!"

She said that in past ages man had lived a life span of hundreds of years on this very planet, when the atmosphere was similar to the one which now exists on Venus. It was not only the atmosphere of Venus, she hastened to add, but the way her people lived, thought and ate, which was responsible for such longevity.

"When we live according to the laws of our Creator we are blessed by the gift of longevity. But this is not the greatest one. It is only a by-product of our blessings."

As she talked she looked up at me, and I suddenly discovered I was taller than she. I was now 25, and we appeared to be the same age.

"When you get over your wonderment," she said happily, "perhaps I can surprise you again. Although you haven't realized it, you have been under constant observation from the moment we parted years ago."

I turned red again, and hung my head.

She laughed.

"No, you haven't always been a good boy. There have been times when . . ." and she made as if she were going to take a whack at me where people usually sit down. I flinched and recovered my composure. Then I laughed with her.

She realized, she said, how my mind and interests had deviated from spiritual teachings they had given to me.

"You have been wrapped up in your business. You have a family—and I can't blame you too much for throwing all your energy into your work. You have also made a great many mistakes in your personal life. But we expect that; no one is perfect. We're not either."

Again, I felt very small. If this wise and lovely creature was not perfect, then what was I.

"We do not condemn anyone for ignorance in such matters. The environment in which you live on this planet forces you to comply with certain social and ethical performances. Man's laws are often good, Howard; but at best they are only misinterpretations of the universal laws."

I asked her what was in store for me in the future.

"It is best that one does not know too much of the future, for it would rob him of the proper impetus to progress and to make decisions. We must learn from mistakes of our past lives."

I wondered what she meant by "past lives."

She picked up my thoughts.

"People live in fear of death, when in truth there is no death. It is only a change from one condition to another."

But she said nothing more about "past lives," and I supposed it was a matter not yet ready for discussion.

"I have good news for you—because I think you like meeting us. You will have many more contacts which will further instruct and condition you. Each contact will be a step in this development. For instance, one contact will deal with "diet"; others will deal with marital problems in a social sense; you will learn a great deal about technology and our science from certain of our people."

She laughed again.

"Yes, you will be busy."

I would also be taught how to develop and use my own mental powers, which she said were present, but lying dormant. I would be able to direct other people through mental contact and assistance.

"You will meet certain individuals of this planet who will come to you at our direction and help you. You will form groups and teach people. Some of these whom you will teach will themselves become teachers and assist you in your mission. One person we will send to you will assist you in forming a body of 12 men to work in conjunction with us, by using the combined thought power of the men in bringing about universal laws and wisdom in the various groups in which they will teach.

"Some of the people in these groups will have contacts in the future, depending upon who can be trusted and who will be worthy of the responsibilities entailed."

Then she gave me a specific direction:

"Howard, your story must not come out until late summer of 1957. Then you must make the story known through all channels of communication, even through some new channels that you would not understand at this time."

She could detect I was thrilled at the prospect of being widely known, and quickly deflated my ego.

"This, Howard, in the way your people speak, will be 'no fun.' Many will believe you and listen to what you say. Even more will resent you and heap ridicule upon you. Not just 'people', but even your own family and close friends, Howard. That will hurt most."

I began thinking. This might not be so pleasant after all!

"You can 'pull out'—now or even later. Are you willing to go on?"

I didn't hesitate for an instant. "Yes," I said.

She leaned forward and kissed me gently on the cheek.

CHAPTER 7

FIELD LOCATION NO. 2

She turned quickly, walked to the ship and stepped into it. The two men followed her. The opening closed and the craft rose vertically. When it was three to four hundred feet above the ground, it disappeared in a westerly direction with a flash of light.

Although it took a few weeks for me to get over the thrill of the new meeting and my mind was not always on my work, I found my sign advertising business growing even more rapidly.

There had been a change in me. Although I always tried to do the best kind of work I knew how to do, I now found myself trying to give my customers a little more for their money, and going out of my way to help them. More and more people heard about my shop and came to have work done. My business continued to prosper.

In the fall of 1947, a young man, neatly dressed in fall clothing, entered the shop.

Though he said he was a real estate man, there was something odd about him. And he didn't act like other extroverted, back-slapping real estate men I knew. He talked for a while about some small matter, possibly the weather, while I could detect he was deciding just how to approach me on some other subject. Finally he told me he was thinking of putting up some "For Sale" signs near a place called Pleasant Grove, about eight miles from the shop, and would like some advice.

It is always pleasant for a man to be asked for advice; besides he was pleasant, spoke softly and called me by my first name from the beginning; so I agreed to go with him.

He did not introduce me to a young lady waiting in the car.

"It's a lovely piece of ground," the young lady volunteered.

"Yes," the man said, "but we feel we are asking too large a price for it . . . in today's market."

The young lady smiled to herself, as if she were enjoying some private joke. I made some small talk about local sales of property.

Halfway there, the conversation fell off and there was silence. Very abruptly the man changed the subject.

"Howard, we know you are keeping your contacts with our Brothers a secret as you have been instructed."

I didn't know if I should feign surprise or not, since I had detected something unusual from the beginning and my suspicions had steadily grown.

"Oh, you ARE. . . ." and I chuckled.

He simply came out with a wide grin, and drove a few hundred yards without speaking further.

"You see, Howard, I have been taught much of real estate, but little of acting."

I could see his message was to be serious, so I simply smiled again and grew pensive.

"We know how it is to hold all this inside you. It is frustrating, for we know how much you have wanted to reveal to everyone what we have told you. You're that kind of person. You would like others to share your joy, your inspiration and enlightenment. That's one reason we selected you."

I felt humble as he spoke to me, softly, directly, always in a very positive manner.

"No, I'm afraid we won't be painting any signs where we're going today. We're going to show you a new contact point. We'll call it . . . say, 'Field Location No. 2.' It is a farm area, quite secluded—a really good place for us to land where no one can be harmed by the electromagnetic force which emanates from our craft."

That was the first time I had heard anything about danger associated with the ships. He sensed the apprehension in my thoughts, and explained.

"Even a small craft, the type we use for observation and reconnaissance will nullify or make neutral anything electrical, such as the electrical system of an automobile, radios, television and the like."

I relaxed, for I had supposed he meant that the forces might hurt people even from a distance. Then I could see his reasoning. Someone traveling near to a landed space craft and finding their motor dead, might become afraid. Especially if at the same time radio and television sets in one area should stop operating. I wondered what would happen if a human touched or got close to one of the machines.

Again he answered my thoughts.

"Howard, we use many kinds of energy which is available to our Infinite Father's children. These powers are all around us in this vast universe. One type of power we use in our small observation craft would kill a man instantly, or over a longer period of time—depending on how high the power was stepped up from the main control switch at the pilot's instrument panel. You'll not understand much of this now, but will remember it later after you are further instructed. The cause of this danger is an electromagnetic flux, or field, revolving around the craft, and the rate of speed of this flux is the determining factor in the amount of damage done.

"Your body is made up of an infinite number of little solar systems, just like your own solar system, each having its own field or gravitic force around it, with particles revolving in their respective orbits, as the planets do. All of these components—and that's a better term than 'particle'—are held in their orbits by this gravitic force, in a relationship of almost perfect equilibrium.

"Think again of your solar system. Can you imagine what would happen to your earth if one of your closest neighbors, say Mars or Venus, or even the Moon, which you call a satellite, were removed from its place in orbit?"

I thought for a moment and, still confused by his talk, was shocked by an imaginary vision of suns and planets spinning around, striking each other, and worlds coming to a horrible end.

"Do you mean to imply that if I touched a space craft when the power is stepped up high enough, it would upset the balance of millions of atoms which make up millions of molecules which in turn make up the cells, organs and so on that make up the human body?"

"I'm afraid your vision of a human body flying apart in all directions is a bit overdrawn," he replied, "but you have the basic idea. The damage this would do to human tissue also might lead to certain diseases which are becoming more prominent on your planet—and a slower death might result, such as cancer of the blood, a malfunction of the liver or some other organ."

The talk was becoming even more interesting to me, and both my companions noticed my enthusiasm and smiled. I immediately enjoyed a wonderful thought that if a body could be destroyed by a power upsetting its natural field of balance, why couldn't it be created by the same great force? Why couldn't a vital organ be cured of a disease once the balance of the millions of atoms and molecular structures of which it is made were set back into natural orbits.

Again he read my thoughts and said, "Howard, we have been curing people with this type of power for centuries. We do not have doctors as you would think of them. Your doctors attempt to cure ailments from the outside: with injections, harmful rays, surgery and other methods. On my planet sickness is a rare occurrence: but when a body does show symptoms of some ailment, this same body realizes that it has been negligent in living one of our Infinite Father's natural laws.

"Improper eating habits, for example.

"When the proper food, grown in perfectly balanced, natural soil, is consumed, it produces healthy blood. The blood is the carrier of nutriment to every part of the body. When the blood is perfect, the body will function perfectly."

"Well, what happens if someone on your planet breaks an arm or leg? What if someone is severely hurt by some unavoidable accident?"

"We are almost there," he said, with the same tactful way of avoiding a subject I was not yet ready to hear about that I noticed in the others I had met.

We rounded a bend.

"Well here we are, Howard: 'Field Location No. 2'."

It seemed strange to me that it had taken so much time to cover the short distance to the place at which we now found ourselves. It had taken at least twenty minutes. I intended to remark about this, but he interrupted my thoughts.

"We did not really intend to make you think we were in the real estate business. People were listening in the apartment above your shop, and as we had pulled up to the curb outside, people were watching us. That is one reason we did not give you our names."

I had often wondered why these people hesitated to give me their names.

"There is nothing mysterious about our hesitance about names. We really don't have any, that is, as you would think of names. I know it puts you at a disadvantage, though, so why don't you call me. . . . (and he thought for a moment). . . . 'L——'—that's a nice name, now isn't it?"

"And what shall my name be?" mused his beautiful companion. They laughed and I could detect that they were finding the process of selecting names for themselves diverting.

"I know, 'T——,'" she said as both she and "L. . . ." extended hands as if in an introduction.

"Very glad to meet you, Howard," "L——" jested.

I shook hands with these wonderful people while an unexplainable and delightful feeling permeated my entire being.

As we walked into the field I perceived it was similar to the place near my parents' home where the first landing had occurred, though more secluded than the former. Evidently it had been "Field Location No. 1."

He turned to me.

"I hope you will be able to remember this location. I realize there are many curves in the route and some confusing turn-offs from the main road similar to this one. But if you will take notice of that tree near the entrance, you will see it is a good landmark."

The tree he indicated was not a large one—about 20 feet high with a trunk about 12 inches thick at the base. But it was the tallest tree in a hedgerow and I was certain I could remember it.

"You will receive a telephone call advising you of the time and place of your next contact. Mental contact can sometimes be confusing, as you well know from your unhappy experience on Okinawa."

So he knew about that, also!

"We'll go back the way we came, Howard, and you can fix the route better in your memory."

We got into the car, and it seemed that in no time we were back at the shop. Again he answered my thoughts.

"We *are* in a hurry. After we leave you we have many miles to travel—deep into the heart of the state of Pennsylvania."

"L——" did not state their business in Pennsylvania, nor did I ask. I was more interested in learning more about the propulsion of their space craft and their methods of curing diseases. These were answers many people would like to know.

Yet, even though I should learn all of this, the information must be kept secret until the summer of 1957. I still had a 10-year period of silence ahead of me. A further period of instruction, no doubt, which would involve more contacts, no telling where or when.

One thing uppermost in my mind was a possible contact where clear, authenticated pictures might be taken, maybe even movies. I wondered if they would permit me.

The car pulled out and I stood watching it leave. About a hundred yards away they hit a "chuck-hole" many strangers failed to avoid, and "L——" looked back at me and waved, at the same time smiling somewhat sheepishly.

CHAPTER 8

THE EXPLODING DISC

The days following the brief meeting with the two inspiring people left me in a physical and mental state which is difficult to explain. I just couldn't get my mind on my work.

My wife noticed a difference in my manner. I would become irritable and moody; at other times I lost myself in deep thought, holding myself almost incommunicado with the world.

It must have been very difficult to live with me some of the time. My mood also affected the business. I got so far behind with my work we suffered financially. Finally I found it necessary to take out a small loan in order to get back on our feet again.

Then as suddenly as I had entered them, I snapped out of my strange doldrums, the business progressed rapidly and all was smooth and calm again. Rose was a loving and devoted wife, and I considered myself fortunate in this earthly venture toward success in business and our marriage.

Our son, Robert; was a great inspiration. His intelligence and physical growth was above average for his age. People were amazed at this precocious child who could walk and utter a few clear words at the age of six months.

One day he astounded a small audience when, at the age of three, he sang "Old Man River" in its entirety. We were proud and happy parents, and once again I was sojourner in this world.

Family life and the sign business went along smoothly for about two years. We had saved a few dollars and bought a small home of our own, one block from the main street in the town of Washington, N.J., with the help of a G.I. loan and the local bank.

We redecorated the house and changed a small garage into a sign shop. We were located in what we thought was a pleasant and peaceful section of town—an interracial, though predominantly colored neighborhood. My wife and I had no resentments nor prejudices toward the Negro race; if we had, we would never have moved there. After living in the neighborhood a few months we discovered that a few of the colored families resented our presence. I learned that other colored families wanted to buy the house we owned, but hadn't been able to manage the down payment. I also learned that the people who sold the house to us were practically forced to abandon it because of neighbors' intolerant attitudes.

They made it so miserable and uncomfortable for them they took a large loss on the property, but still were happy to move.

We realized that as a minority family we would be subjected to the same intolerance, but we still tried to avoid forming any resentment toward our neighbors. We believed that all races and creeds could live together in peace and harmony. In spite of our own attitudes of tolerance and good will, the situation worsened.

Whenever I left the house on business to see a potential customer, my family suffered all kinds of abuse and indignities. One time I came home early and found Robert, then six years old, surrounded by about 15 children, most of them over 10. They were pummeling him with rocks, some of which struck him on the head. They were circling him with rough-hewn spears and keeping in time to a jungle-like chant accompanied by a home-made drum.

Of course I saw "red" when confronted with such ignorance, regardless of the color, race or creed involved. If it had been childish roughhouse, I could have forgiven it, but I had spoken to the parents of the children time and again, asking them to make their children halt such unmannerly treatment of my own young children—but it was to no avail. We couldn't let our children walk to school alone, fearing they would come home beaten and battered. We learned that the school was also having its share of difficulties with the same unruly children. My wife found it unsafe to be seen outside the house in our own back yard. She would be insulted, abused, and many times she had to retreat back into the house under fire of rocks. Their plan was working, as it had with the previous occupants, and we made up our minds to sell at a loss, also, and make a search for a more peaceful community which would be more conducive to a successful sign business and a happy home life.

There are many reasons why I have decided to make this particular incident known to the many wonderful people who will read this part of my life, which was not a pleasant part of it. One reason is that from the very beginning of history anyone, or any group or nation, who has attempted a sincere move toward living in a peaceful coexistence with others has failed because one side or the other has reached a point where jealousy, greed, resentment, or misunderstanding has come between their difference of color, race or creed. This has usually ended in hatred. Sometimes it has resulted in physical as well as mental pain—in torture, murder—and then we must have our wars, whereby we attempt to solve the unaccepted differences with force.

What happens in a small neighborhood, in towns, cities, states or nations is summed up by the same ugly nine-letter word: IGNORANCE!

The feelings toward us in our neighborhood were not all bad. One wonderful fellow, also in business for himself, expressed himself to me in almost these same words. He still lives near there. He is colored. But isn't it a shame that we should identify a man by his color, rather than his deeds? I believe a man is entitled to freedom and peace of mind, regardless of his color or religion.

All of us have a spark of the Infinite Creator within us, and certainly this spark is of no different color in different individuals. The soul is beautiful under the skin, regardless of its physical shell in which it is expressed through natural law. There are stages of evolvement, but one is not any "better" than another.

Our children who are growing up in this world of ignorance won't change the world into a better one of understanding and love, unless we, the parents, along with the schools, colleges and churches, teach them—not with words but with deeds.

In our churches we teach our children to be good else God will punish them. This, in my understanding, is a degree of ignorance. GOD DOES NOT PUNISH ANYONE. Anything which happens to us we bring upon ourselves. God is all men, all things, and is present everywhere, in an Infinite Universe. Man is but an infinitesimal, physical, limited expression in this three dimensional world. Man is an insignificant frog in a barrel, who has no concept of the beautiful lake on the outside of his little world.

But I must get back to the chain of events which culminated in even stranger sights and more amazing revelations to me.

In the spring of 1950 we visited my borther-in-law, who told me of an interesting experience shared by his wife while they were driving in the country a few weeks before.

They were driving slowly through Pleasant Grove, off Rt. 24, a few miles from their home, when a blinding, orange flash attracted their attention. They slowed down and watched the fiery ball explode in mid-air over a field to their left. The entire field below was lighted up by the explosion and appeared to be on fire. Becoming frightened, they drove away as quickly as possible.

I listened with interest and asked him to drive me to the place. We got into the car, and I noticed we were approaching what I called Field Location No. 2, where I met "L——" and his friend in the fall of 1947. He stopped the car. There was the tree we had established as a landmark. There could be no mistake. This was the same place!

I grew excited, though I tried to mask my enthusiasm. We examined the field. It had been plowed up since the explosion; that could have obliterated most of the signs. I looked at the largest tree again. It had been burned slightly, and one of the branches had been broken.

Thinking to myself, I guessed the damage might have been caused by an observation disc sent out by a control ship. Perhaps the small disc had got out of control and was exploded before it could do any serious damage. Perhaps it had been heading directly toward the car when it was harmlessly destroyed.

As we drove home my brother-in-law wanted to discuss the incident further, along with various subjects relating to space travel and the possibility that other planets were inhabited.

As I talked with him I realized that many people were growing interested in such previously "ridiculous" subjects. Again I could see the great wisdom of the space people. Their asking me to wait before I told my story was wise indeed. Perhaps they were waiting until an increasing awareness in people made acceptance of my story more assured.

I felt my way carefully in the conversation, all the while noting his reactions to different concepts I put forth. I wanted to tell him much, much more, but I could not allow him to suspect at that time my own part of and work in this field. Again I found it extremely difficult to know a wonderful secret and be unable to share it with others.

In June of the same year I was working in the shop when I received a strong telepathic impression—but this time it was different. I had the distinct sense of hearing something like a telephone or distant radio inside my head. It was a man's voice, and it sounded like "L——."

"Howard, I am waiting for you in a car at Field Location No. 2," the voice said. "Please come as soon as you can."

I was thrilled at the prospect of seeing "L——" again. As I hastily cleaned my brushes I wondered how it was possible for "L——" to communicate in that manner, for I had heard the voice distinctly, even though the location was about nine miles from the shop. I put aside my work, got into the truck and headed for Pleasant Grove.

As I approached the dirt road leading into the field, I passed the tree landmark and saw a car parked farther down the field near the trees which separated it from another field. I parked the truck and approached the car.

"L——" got out of the car and we shook hands warmly. He apologized for taking me away from my work. He said there were many ships in the surrounding area of New Jersey, New York and Pennsylvania, and there were as many people from other worlds working with the people on this planet, just as he was with me.

"This is only the beginning, Howard. There will be courses of instruction and processing. In fact you will find it hard to believe some of the things you will be doing in the future, but you will be properly prepared so that you will be able to cope with these unusual tasks."

I mentioned the exploding disc, and he confirmed it to have been one of their observation discs out of control. They had been operating in the area, he told me. Many people had seen the explosion, but had passed it off as a natural phenomenon.

"I'm sorry the tree was damaged, but the limb will be healed and it will continue to grow," he promised.

I thought if these wonderful people were concerned about a lowly tree, how much more must they be concerned with their fellow men!

Many more sightings would be reported in this area, he told me, as well as other places in New Jersey. That fall there would be a sighting over Washington, N.J., which would be witnessed by my friends and neighbors.

The experiences my family was going through were for a purpose, he said. He understood the situation clearly and indicated that changes would occur sooner than expected.

I asked him when we would meet again, but he said he did not know at that time. I would have other contacts, he promised, who would instruct and guide me in carrying out my work. He suggested I no longer discuss the incident of the exploding disc with the family.

As we left the area, I headed west toward home and he headed east. As I drove home, all the questions I wanted to have settled came to mind. That night I could not sleep, as I thought and wondered. Who were these people? Where did they come from? Why were they here? Why did they contact me? How were they able to communicate over long distances, from one mind to another? What method of propulsion did they use? Questions . . . questions. These and many other thoughts crowded out sleep.

CHAPTER 9

STRANGE INSTRUMENTS

In August, 1953, I again drove up to Field Location No. 1, and there, as I expected, was a car—a 1953 Chevrolet. It was a red and white convertible, with Pennsylvania license plates.

I locked my car, walked over to the convertible and got in at the invitation of two men.

Something about the men was different, though they were of average height, had brown hair and dark eyes, and were dressed in business suits. They must have noticed my unintended stares, if only from the corners of my eyes.

Then they eased my curiosity. "We happen to be Martians," they said.

I inspected the car further. Brief cases, maps and papers covered the back seats. Strange instruments on the floor then attracted my curiosity. They were box-like objects, about the size and shape of shoe boxes, apparently made from a plastic material. A coil, which reminded me of an antenna, protruded from each of the three objects of different size. All were the same color, a soft grayish green.

I immediately guessed (and correctly as I later found out) that the color was for camouflage, and the objects were to be spotted in fields, bushes, and so on.

We pulled out of the field and headed toward Easton, Pa., and as we rode away we talked about many things. I was hardly prepared for the unusual things I was to learn.

Previous contacts had mentioned I was not the only one working with them. Now they further confirmed this, giving me actual names of people I eventually met and recognized. Not only were they contacting people in the east, but other sections of the country as well: I remember they specifically mentioned California, New Mexico and Arizona.

Not always did people know they were working with the space people, even though their contributions had been valuable.

"Two men have been working with us for many years, rather indirectly, and, Howard, they don't even know it. They're being directed, but they don't know it, let alone realizing who is helping them."

Three other men they mentioned had been working

independently on scientific projects and were finally contacted because of their work.

Then they began to inform me of some of my own future work. My work would mushroom into a much larger scope and would take on world-wide significance. But at that time I never in my wildest dreams thought I would be making cross-country lecture tours, appearing on radio and television, meeting and talking to large numbers of the public.

Neither could I foresee the opposition and persecution I would encounter almost every step of the way.

As they remarked of future happenings, again I learned of the long series of instruction and training I would receive. Some of this would be immediately practical. They knew I had been a topographical draftsman in the army, but nevertheless I had a lot to learn, particularly in certain phases of mathematics such as trigonometry. It would be necessary that I be able to estimate the distance of a mountain or the height of one of their craft, for example.

The reader likely can understand why the sign-painting business became less and less interesting, as I wanted to devote more time to this fascinating, exhilarating and humane work. More and more it was becoming difficult to fit my mind and aspirations into the cubicle of one small community after I had glimpses into the universe, its many worlds and peoples.

When we arrived in Easton they pulled up at one of the restaurants at the circle in midtown.

I noticed neither man ordered meat, but asked for vegetable platters and fruit juice. They had coffee which they took black. Feeling they were engaging in some sort of religious fast, I ordered fish, which I didn't feel would be inappropriate.

Sensing I was slightly ill at ease, one of them explained their diet.

"Don't think of this in the usual sense of a religious rule," he told me, "but of natural law."

He continued:

"The dietary customs of your various religious groups often approach, but are still misinterpretations of these natural laws. Rules of diet should never be thought of as sacrifices, but as positive contributions to health and the exercise of consideration toward lower animals.

"Your own diet, Howard, as the many other things you may or may not do, will be up to you. And we don't expect you to change overnight."

They suggested that if one wished to change from meat-eating habits to a more humane and better diet, one should begin by eliminating first the blood-red meats, such as beef, pork, lamb, and so on. One should then turn to fish and fowl. Next, fowl should be eliminated, and finally fish.

"If you still feel the need of having animal protein, fish is preferable due to the lower consciousness of the animal. In time you will discover vegetables with such high protein values that you can eliminate meat altogether—as we have done on our planet."

Then they talked further of their many contacts all over the world and my job in particular.

One of my tasks was the mental assistance of individuals, often without their knowledge. Such could be accomplished by sound-frequency waves, light waves, the use of colors and other physical means. I had always thought of such matters as being accomplished in some supernatural manner; but I was rapidly learning that the Infinite Creator accomplished all purposes by natural laws.

"Do not think of this as some artificial control of the human brain," one of the men said, "as you may see in some of those horrifying science-fiction pictures—though I must confess that he (and he indicated the other man) and I saw two of them on a double picture (I assume he meant to say 'double feature') and rather enjoyed them, if mainly for the humor we saw in them. We do not control the brain. Such an action is not in keeping with the laws of the Infinite Creator. Instead, with the proper instruments and technique, you can accomplish a much larger purpose: YOU CAN RELEASE SOMETHING IN THE BRAIN WHICH IS ALREADY THERE."

The instruments in the car accomplished such tasks. I would be asked to place them in four states: New Jersey, New York, Pennsylvania and Maryland. With each instrument would be a man or woman who would act as a human terminal in conjunction with the machine. These people would react according to their own individual brain development with the assistance of the impulses received by the machines.

I didn't quite understand how it was to work, but it seemed they were trying to get across that it was necessary to have both the machine and the human mind working in conjunction to accomplish such a purpose.

"A central location will be established for each instrument, The range of each instrument is about 25 miles.

"After you place these instruments, Howard, you will notice an immediate and obvious effect. People within a radius of 25 miles of each instrument will automatically become more aware of and conscious of their interest in space travel and in our space craft. They will see more of our craft because they will be looking up. Then when these people hear of your experiences, they will be inspired to come to you and offer to help you in any way they can."

They had already established many bases. The one in New Jersey was, they said, within 15 miles of my home, totally unknown to anyone, except, of course, those of this planet who had been working with the space people for many years.

It seemed these Martians spoke more solemnly than the others I had met, without the many flashes of humor and good nature the others had exhibited. But perhaps it was because they weren't considering me as a child any longer, and knew it would be only a short time before I must accomplish the biggest part of my task: informing others of my experiences.

I was bursting to tell the world. At the same time I felt a tremendous responsibility to these people from space, as well as those on our own planet. They made me feel as if so much depended upon my skill in accomplishing some of their purposes. And I'm afraid my ego was often inflated to a great degree.

Soon would come a time when all my ego would vanish as it would be quickly deflated by the continual abuse and ridicule I would suffer.

"You will be warned, and threatened to stop," they admonished me as they dropped me off at my home.

And again, the question, "Do you still want to go on?" And my answer, as always, "Yes."

The next few years were uneventful, so far as contacts were concerned, though I was given a course of instruction which I do not care to divulge at this time. Because of the increase in the size of my family and added financial responsibilities, I spent most of my time working in my business, and made a comfortable living.

I was to experience a period of personal loss and unhappiness. In 1954 my oldest son, Robert became ill. He had a brain tumor and he later developed cancer. Rose was a patient and loving nurse, completely devoted to him.

In September, 1955, my younger brother, Alton, was

killed in an automobile accident. A few weeks later, my mother left this earth.

Gradual changes and upheavals in my family were to have subtle effects upon the overall picture, and with them, I believe, I learned many lessons, particularly of humbleness and the insignificance of man. During this period I was thinking constantly of the better world which would come about, were the tasks set before man by the space people performed with dispatch and zeal.

My father was now very lonely, and in October, 1955, he asked us to move back to the old home place.

There began a series of events which would bring world-wide attention to High Bridge.

CHAPTER 10

BARBER TO THE SPACE PEOPLE

In March, 1956, the telepathic invitations to contacts started again.

The nature of the meetings required that many of them take place at night. Often I would receive such mental impressions between 1:00 and 2:00 A.M., and, while my wife lay sleeping, I would drive away to meet the space people and be given further instructions pertaining to my work.

Since I was working by day, and often up much by night, sleep and rest became a premium and a luxury.

Many of my contacts were of strictly mundane nature, however. I found I was actually helping them in little material ways, and such occasions I enjoyed as much as the periods of instruction.

Often I purchased clothing and took it to the points of contact. Visitors just arriving from other planets had to be attired in terrestrial clothing so they could pass unnoticed among people.

Such tasks were not without their moments of humor, and I think the visitors enjoyed them as much as I did. I remember one time when I was asked to purchase several complete outfits of female clothing. Feeling it would be embarrassing and somewhat difficult to explain why I was buying so many outfits, I purchased them in separate shops.

I bought what I thought was the appropriate sizes and showed up at the point of contact. The women went into the next room from which I soon heard a series of giggles and groans. Finally the door opened and the bras were flung out. They apologized, saying they just could not wear them, and they never had. Just why I didn't know, and you may be certain that I felt it wise not to ask!

When they had put the rest of the clothes on, they marched out and walked up and down, more amused at their new finery than proud—though I must say that they exhibited natural feminine instincts in being excited about the new clothing.

Once again they had difficulties—the high heels. They teetered and wobbled and suffered, but took it in good humor. They realized they would have to learn how to wear them, though they often complained, "Why can't your women wear sensible shoes!"

I ran into equally unfamiliar tasks in helping the new arrivals. A man with long blond hair, which hung to his shoulders, approached me and handed me a pair of scissors. He could not yet speak our language, as many of them couldn't at their first arrival—though they would soon learn it, and speak it fluently.

This man simply pointed to his hair and sat down, and I presumed rightly that I had unwillingly become a barber!

I took a handful of the finely textured hair and opened the scissors. I halted and looked at the man. His plaintive look made me feel sorry for him, for it was obvious he was proud of his beautiful hair. But at the same time it was grotesquely funny. He laughed and motioned to go through with it.

I remember several occasions when I cut their hair. I don't know if they save the hair or not; however, all evidence of the meetings was always carefully gathered up by the space people before they departed.

The men, particularly the Venusians, had unusually fair skin, without hair on their arms or face, and had no need to shave. After three months on Earth, however, they became hairy and grew beards. Most of them waited the three-month period so they would have beards to shave and appear more like earth people.

Some of them requested dark glasses: a few asked for dark glasses with red glass, which was difficult to obtain. I don't know why they wanted dark glasses, for those I had met previously had not worn them.

There were so many things to ask about I never remembered to inquire, though I presumed the glasses were used to acclimate their eyes to the glare of the sunlight here, which probably is more intense than the light on their planets.

Thus I had the opportunity to meet people from other planets in all stages of progress and devlopment: from those who spoke no word of our language to those who spoke it fluently; scientists and technicians to helpers and assistants.

I briefed them on our customs, slang and habits. Although they utilized instruments to learn a language quickly, the machines couldn't always cope with colloquialisms. And they had to pass for ordinary people. I was gratefully surprised to be of immediate help to these advanced people.

Occasionally they asked me to bring food. They asked mainly for frozen fruit juices, canned fruit and vegetables, whole wheat bread, wheat germ and the like. They refused to drink milk, avoided fresh oranges, lemons and grapefruit. They preferred tree-ripened fruit when I could find it; when I couldn't they had me obtain frozen food from supermarkets.

I remember one time I bought five bushels of tree-ripened apples from a local orchard. They tested the apples and found the mineral and vitamin content much lower than similar fruits of their planets. This was due, they said, to our poor soil. They explained that chemical fertilizers were not the correct answer to the problem of our impoverished soil, because they did not replenish the organic materials our soil sadly lacks.

Most of the time they brought their own food, in a dried, preserved state. I sampled some of it and while it was tasty, it was hard and dry. I tasted other food which was delicious. They had put the dry foods through some sort of processing which returned the moisture and at the same time expanded it to its natural size and state. I ate one of their tubers which was far superior in protein and mineral content than any vegetable we now have. We could grow the same here, they said, if our soil were healthy. Their preserved foodstuff would not spoil, mold or decay, presumably would last indefinitely, ready to process and eat at any time.

They never asked me to obtain any identification papers for them or to help them locate jobs. They seemed to be able to take care of such matters themselves after they had been properly acclimated and grown accustomed to our ways. Once clothed in our attire and briefed thoroughly in our customs, they were on their own, and seemed to experience no difficulties.

Often I would stay with newly arrived visitors several days, helping them adjust and become familiar with the locale or town in which they were staying. I would gather information for them, such as where the post office was located, where the schools were, the source of water supply, whether there were bodies of water nearby, and so on.

I thoroughly enjoyed this work, and while it did consume a great deal of my time and money, it was a pleasure; and as I briefed my friends I learned more and more from them.

CHAPTER 11

THE OBSERVATION DISC

On a warm, spring night between 10:00 and 11:00 P.M. in April, 1956, just after my dad, who had night employment, left for work, I received a strong impression to go up to Field Location No. 1.

To approach this hill, which is at the rear of the old home place, one must drive around the other side, where there is a plow trail leading up to the field.

It was a dark night, but since I did not want to attract attention, I drove into the field for about a thousand yards with my car lights turned off. I arrived just in time to see a craft coming in over the crest of the hill from the west. I leaped from the car, at the same time grabbing my candid camera. I snapped a picture of it.

The ship took the form of a pulsating, fluorescent light, changing in colors from white to green to red. As it neared I prepared to take more pictures. It came in slowly, at about the speed of a Piper Cub. When it was within a foot of the ground and about a hundred feet from the car, it hovered, and I recognized the familiar bell shape. The pulsating colors stopped, it gave off an eery, bluish light, and then portholes appeared.

It lighted the field immediately around it with a soft, subdued light, as I snapped more pictures before it landed. After it touched ground, a man appeared in front of it, and I quickly photographed his form against the backdrop of the craft. Then I walked in his direction, and he acknowledged my presence by beckoning to me.

I put the camera in my jacket, walked to him and shook hands with a tall, handsome man whose blond hair fell to his shoulders and who wore a one-piece uniform. He led me across the field, and when we were at the edge of the woods he pointed to something lying on the ground among the trees.

It appeared to be a circular, translucent object about a foot thick and three or four feet in diameter. It was pulsating different warm colors. As we approached it, the colors changed from white to blue and back to white with a tinge of yellow.

"Hold it, Howard," my friend warned, as he put his hand on my shoulder. "This is as close as we can go safely."

I noticed his voice was rich and deep as he informed me it was an observation disc controlled remotely from the ship to our rear.

"This little disc is recording all your emotions, thoughts, possible intents."

I didn't feel a thing, though it was shaking to know that a mechanical device was reading my thoughts.

"Don't worry," he reassured, "it's white. When it changed to white I knew you were all right."

He said the same colors were being indicated on an instrument panel within the ship and would be recorded on a chart which would be kept as a permanent record.

"I want you to get a good look at this thing, for you'll have to bring people up here before they'll believe you. We'll stage a 'test' for them."

He exhorted me to inform such future witnesses they must stay at least 20 feet away from the discs because of a certain amount of field radiation when the instruments were transmitting and recording.

"We'll use this location as the site of an experiment. If it works out all right here, perhaps we can get some idea of what would happen if people saw this over an entire state or even the whole country. This is one thing that is strictly experimental and we have no idea of the results. We know a lot about your people, Howard, but figuring beforehand just how they will react to this is impossible even to us."

I wondered why for the first time I had experienced an urge to photograph their craft, and for a moment had misgivings, feeling he might not approve.

"Your pictures will also be a test. Eventually you are to show them around. Tell people we're from other planets and that we are not hostile. We realize we cannot convince all the people of this world at one time; it might not be a good idea, anyhow; it might come as a great shock to people in low states of evolvement."

"The pictures should convince ANYONE," I rejoined.

"No, you're wrong, not everyone. The radiation of our ships will make them fuzzy, and that will do no harm. As I said, we don't expect to convince everyone immediately, nor do we want to. It will of necessity be a slow procedure."

Again I was reminded that the story should not come out until the summer of 1957, and I knew he had also learned our language well, for I remember how he put it in slang:

"Don't jump the gun, fellow."

Should it come out before that time, I would be subjected

to even a greater amount of ridicule and persecution by those who would not understand.

"Goodness knows it will be difficult enough for you, Howard, even after we've got you ready for it. You must know how to present the story properly, and not spring it all at once. People just wouldn't believe you if you went about it in the wrong way and more harm would be done than good."

I asked him how I should go about choosing the witnesses and was told that I would exercise my judgment and my judgment would be good. Some people would come to me of their own accord; some I would choose.

I asked him about a close friend of mine, Bill Thompson, in whom I placed great trust.

"Oh, Bill. We know him. In fact we tried to contact him in 1933!"

Bill would believe and understand, he said, though I was not to confide everything to him. I was glad to hear this, because ever since we were in high school together we had enjoyed a close affinity. I would be extremely happy to tell at least someone of my experiences.

Our conversation was interrupted as the observation disc suddenly rose, becoming brighter as it gained altitude, and lighting up the woods beneath it as it projected patches of light and dark through the trees. It rose to about a hundred feet and then went in the direction of the ship where it seemed to disappear within the craft.

As we walked back my friend explained such discs always return to the point from where they are sent out from ships on magnetic beams similar in a way to a bowling alley where the ball returns to the bowler in a grooved channel.

"You've just had a going over that would put any psychiatrist to shame," he remarked with a trace of a laugh in his voice.

"Is it something like psychoanalysis?" I asked.

"Yes, you could say that. But it is far superior. Psychiatrists touch only surface conditions, while these probes go much deeper into individuals' makeups. In fact while that disc was there near you it gathered an entire summation of all the lives of your particular soul. But there is one thing it can't tell us," and with the last remark he was pensive.

"What is that?"

"Man has free will; the pattern of his emotions and thoughts and consequent future actions are subject to change. No, Howard, our pretty little light is not infallible."

Near the ship we halted. I desperately wanted to go inside, but managed to control my desire to ask. Instead I surreptitiously tried to send a strong mental impression to him indicating my wishes, while trying not to show eagerness in my face.

He smiled at my thoughts and told me they did not have the time now and that I would need certain processing before entering one of the ships. Often at parting I had the desire to go with them and leave this planet, though I realized that was not their plan for me.

We shook hands and as usual I was advised I would be contacted again very soon.

As he prepared to enter, he told me to step back at least 50 feet, and I complied. Once he was inside, the craft rose vertically, and when, as usual, it was about a hundred feet off the ground, there was a flash of light and it disappeared

While returning home I wondered how I would be able to continue keeping secret all of the wonderful things which were happening to me. Surely someone must have seen the lights during my many contacts. Someone could have been in the woods and seen the whole thing. I could be certain, however, that only one witness besides myself had seen it all—and that was my dog, Tassy, which had been with us all the time.

I remember how at one time my friend bent down and petted Tassy and said she was a fine dog. Now, Tassy, however, had taken this all in stride, and had run off through the bushes after a rabbit.

CHAPTER 12

THE STORY LEAKS OUT

I sent my film to the processor with misgivings, fearing someone would notice what was on it and grow curious. I had thought of having a friend do the work, but he would be even more likely to look the pictures over closely.

I somewhat breathlessly waited for the prints to be returned. But my fears were groundless and I was somewhat disappointed to see that only one picture was fairly good—the shot showing the man in front of the craft. The prints showed very little detail. The shape of the craft was greatly distorted, and the man showed only as a dark silhouette without the contour of his fine body showing through the uniform: instead of being beautiful as I had hoped, the picture was grotesque!

Then a dreadful thought went through my mind and I chilled. These pictures must have been duplicated and faked to resemble the pictures I had taken. Though in spite of the fact that I sincerely believed the pictures had been cleverly faked duplicates of the originals, refusing to remember, in my disappointment, what my friend had said about the field radiation around the craft, I nevertheless showed them to my family and a few friends.

The next day I went to the small processing lab and asked the owner if he had made up the pictures. He admitted he had not. He had been very busy and had farmed them out, along with some other work, to a friend. I was still quite suspicious, yet I made no accusations.

My real point of frustration came when I tried to convince my father of what I had seen. In spite of the visual proof I offered, which, of course, wasn't good, he remained skeptical. Because some people thought the pictures had been faked, and to conquer my own misgivings about the photographer, I decided to buy a Polaroid camera which would develop and print pictures within a minute, right inside it.

One person in whom I confided believed me. That was Bill. It was indeed gratifying to make some progress at what seemed to be a hopeless and discouraging task.

I purchased a second-hand Polaroid camera and patiently awaited another opportunity to take more pictures. I did not have to wait very long; for on a night shortly thereafter I again received a telepathic communication to come to Field Location No. 1.

It was 12:45 A.M., August 2, 1956—a sultry summer night. The heavy air was pierced by the chirping of crickets as I stood by my car with the camera poised, looking upward and hoping for a chance at some good pictures. Surely enough, over the west side of the field I soon saw a craft sliding noiselessly through the dark sky in an easterly direction, appearing as a bluish white blob of light. The closer it approached the brighter it became. Finally, when it was within a hundred feet of me, it slowly descended and hovered about two feet from the ground.

I snapped away, hardly able to wait for even a minute while the pictures developed, but in the darkness could not see just how I was doing.

I noticed one of the three ball-like objects under the craft became distorted and looked like rubber as it seemed to stretch and grasp the ground. I could see the other two balls through the translucent flange. I wondered how they could make metal appear translucent, and also become plastic, certainly alien to our earthly physics.

Portholes in sets of three then appeared around the dome. An opening appeared and a man stepped out. He stood tall and straight, his long, blond hair blowing in the soft warm summer breeze. I could see the beautiful structure of his body: his broad shoulders, slim waist, and long, straight legs. His ski-type uniform covered his entire body, exposing only his head and hands. He approached and when he was about 50 feet away I snapped his picture. He, like the other man, was silhouetted against the glowing craft, a dramatic pose which I hoped would turn out better than my previous picture. But in the picture the craft seemed distorted and looked as if a gaseous, swirling haze encircled it.

As I had raised the camera he had paused; then he walked forward to meet me and as we greeted each other again I had that pleasant feeling of warmth and love coming from him to me.

"I hope these pictures will help you in the future, even though they may be slightly distorted, due to the electromagnetic flux around the craft."

I then realized the first pictures had not been faked and I felt ashamed of how I approached the photographer so suspiciously.

"It is not the fault of the film nor the developing process," he added; "it's just that the film doesn't see things exactly the same way your eyes pick them up."

He suggested that I try to get a good shot of the craft taking off when I left, and promised that I could take more pictures the following night at Field Location No. 2.

He told me I would receive either a telepathic message or a direct telephone call, but in the event of the latter, gave me a code word to identify the contact so I would know it wasn't a hoax.

"Your story is getting around, so prepare yourself for pranksters," he admonished.

At the same time he gave me names of possible witnesses who would come into the plan. One was a student physicist studying in Princeton University who had been under their observations for a number of years. The other was a friend of this man, who lived in Washington, New Jersey, and who was described to me as a brilliant high school student. Both individuals, he said, were highly evolved people who conceivably could help if it were their desires.

Since there was to be no force or persuasion attached to the work, it would be on a volunteer basis and completely of their own volition. The high school student would be instrumental in introducing me to the university student and both probably would be present as witnesses in the near future.

My contacts would now be speeded up, and I would have many witnesses. Hundreds of people would come to me and I would be much in the public eye. Nonetheless I was counselled to exercise wisdom and perfect control over situations which would come up.

He also told me that within a short time I would meet a very highly evolved being—a great instructor from one of their advanced civilizations on another planet.

Then he changed the subject. He knew I had been worried about how I had revealed some of the information I should have withheld.

"You are inexperienced, and, anyway, the proper time for revealing it is almost here—so little harm has been done."

He wished me luck and stepped back into the craft. As it took off I was able to get two successful shots of it.

As I went home I was lost in a sea of thoughts. Although breaking the story would further explain my unusual hours to my family, how could I get this story across to the people. Why was I chosen to do this? I felt inadequate, helpless.

The story had jumped the gun by eight months because

my wife had become perturbed about my getting up and leaving at strange hours of the night. To ease her concern I finally had to show her and the immediate family the pictures I had taken, and then, of course, tell them something of what I was doing.

Thus, due to conditions perhaps beyond my control, the story did get out before the appointed time. This worried me, for I had been forewarned of the repercussions which lay ahead.

CHAPTER 13

ANTI-MAGNETIC FIELD

This should be one of the most interesting chapters in this book, but in reading over the rough draft I made of it (I have to write something three or four times before I'm able to get most of the grammatical kinks out of it, and then it is not great literature as you have found out), I found it was impossible to describe the feelings I had when I first entered one of the space craft. Somehow, when that great moment arrived, the overwhelming stimulus made my mind flee from it and instead of looking over the ship, as I should have, I seemed to develop, instead, a fixation about why my watch had stopped.

In fact, it was almost as unnerving as my first television appearance.

But to get the cart behind the horse, I should say it was about midnight the following night when I received the telephone call. A man's voice first gave the code we had agreed upon and advised me to get up to Field Location No. 1 as soon as possible. Before hanging up he said there would be another call on the Fourth of August.

I arrived at the field just in time to see a small space craft descending slowly. The brightness diminished and the color changed to a deep blue violet and then almost disappeared. It landed about 50 feet from me, and for the first time I heard a sound when one of these ships hit the ground.

On the ground it appeared as a dark hulk with no reflecting lights on it—only a soft, bluish, white light emanating from the portholes. From one side of the craft a brilliant light shot out and I could see the outline of a tall man in the opening. He emerged from the craft and walked toward me as another man appeared in the opening, waiting.

As the first man reached my side he greeted me, "Howard, did you get a good picture?"

I had been so fascinated by the landing I had simply forgotten to take pictures; so, embarrassed, I grabbed the camera and snapped some shots.

Apparently they were trying to cooperate with me and were anxious that I get some good clear pictures, realizing how difficult it was to photograph the craft.

Then I received a tremendous surprise.

In a very matter-of-fact manner the man remarked, "We don't have much time, Howard, so come on and we'll make a quick trip to Field Location No. 2."

He turned, as if to indicate I should follow him to the craft. In my surprise I just stood there, with, I am afraid, a wild stare of wonderment in my eyes.

"Just a second," he said, as if I had already started to walk with him, and he motioned to the second man, who stepped outside and raised his arm. In his hand was some sort of instrument which he pointed at me.

Suddenly a bluish beam shot out at me, and as it struck my head I felt a tingling sensation, warm, and rather pleasant. I stood in my tracks as he slowly played the beam downward over my body until it had reached my feet. Then he turned it off, and my friend threw up his arm, indicating that I walk ahead of him. The other man had stepped into the craft and now beckoned me.

I walked up to the ship and through the opening, which closed behind us. I was almost stunned. Surrounded by the strangeness of the inside, my attention must have simply retreated to the familiar, for I remember looking at my watch and regarding the face of it in an inordinately close manner. The watch said 12:45.

It had stopped, for the second hand was not moving. I shook it and put it to my ear. Then the hand began to move and I could hear ticking.

The man who had stayed with the ship was at a strange instrument panel. I noticed he was watching my performance with the watch with some amusement, for he grinned. At that moment the walls became brighter, as if illuminated somehow from inside themselves, and I felt a sudden jerk. I assumed we had left the ground, though I could feel no movement. One moment later I felt another jerk and the opening appeared again in the wall.

"Here we are," said the man who had come out to meet me.

My friend stepped out first and I followed. Again I was amazed. There, etched against the sky, was the tree landmark I had grown to know so well. We were at Field Location No. 2!

The craft was now flat down on the ground, dark and obscure, and probably impossible to see from any distance. The night was very dark, and the only light was a beacon, about five miles away, which flashed on and off. The neighboring farm houses were dark and silent, and the only noise was the intermittent barking of some farmer's dog in the far distance.

Recovering from my surprise and wonderment at my first ride, I first asked about the bluish beam they had projected on me; then why the jerk when we took off; why my watch had stopped. Then almost before they could answer, my pent-up mind let go with a dozen more questions.

My friends just laughed, and one of them said they would try to answer all my questions if I would give them time.

The trip, he explained, was an experimental one, to determine if it were possible to take pictures of the craft within its flux area, then get into the craft with the camera, then take more pictures without ill effects upon the film.

I was flattered when they explained they had rigged this craft especially for me, having installed special accessory instruments which they hoped would prevent the magnetizing effects and the radiation which often ruined films, magnetized our watches and affected our electrical equipment.

I mentioned I thought my watch had stopped when we entered the craft, and he said it had indeed. When the man at the instrument panel had pushed a certain button, however, the magnetic condition had been nullified. That was why my watch started again.

He hoped the new instruments would remedy the magnetic conditions which affected cameras and other equipment and possibly could do physical harm.

"Now for your first question," he continued. "We projected the beam on you to condition and process your body quickly so you could enter the craft. What actually happened was that the beam changed your body frequency to equal that of the craft. Thus you felt entirely comfortable inside the craft and suffered no ill effects."

"Tell me, Howard," the other man asked, "Did you have a burning sensation when the beam hit you?"

"No," I replied. "It felt warm, and it rather felt good."

"Excellent," said the other. "Had this beam hit you before you knew of us and our craft, your fear would have caused a chemical reaction with the ray which would have in turn caused a hot or burning sensation. It wouldn't have hurt you, but it would have been mighty uncomfortable."

I then understood one of the reasons why they approached earth people so carefully. Fear must be one of the emotions they must eliminate before they can safely approach us. Fear probably could cause an earth man harm. Then I understood a good definition of fear. Fear was the absence of love and trust.

When they apparently felt they had answered enough questions, they suggested I try more pictures of the craft.

I felt they were actually enjoying the picture-taking and that their insistence about it was made half for their own enjoyment. As they walked eagerly to the craft, entered and took it off the ground, I was reminded of my own small son's eagerness when we often took pictures at home.

The ship rose about 25 feet off the ground and headed in a westerly direction for about a thousand yards. I snapped several pictures of it, and when they returned I showed them the prints. They examined them carefully, as one of them reached for a little brush which comes with the film and rubbed the picture with it to preserve it. It was evident that their new equipment had been successful, for the film had not been damaged. But they shook their heads: the pictures were still not very clear, and they said their equipment needed further adjustments.

I wondered why all the fuss about my taking pictures when they, themselves, must have far better cameras. I asked about it. The two men looked at each other and smiled.

One of them reached somewhere inside the ski-type uniform, though I could see no pocket nor opening, and withdrew something shiny which he held in the palm of his hand. He then opened his hand slowly and indicated I should look.

As quickly as I had seen a flash of what he held, he closed his hand and put the thing back inside his uniform, presumably into some pocket. In a brief instant I had regarded what for loss of words I shall call a picture, but which is impossible to describe. It was of some scene with much green and blue sky and fantastic buildings. It was more than a picture. It was self-illuminated and was like looking more through a window at a real three-dimensional scene than at a flat object.

"You wouldn't want to scare them to death, now, would you, Howard. What would your grandfather have said had he walked suddenly into a room where a television set was turned on?"

I didn't ask any further questions about their cameras.

They indicated we should leave, and in another unbelievably short period we were back at Field Location No. 1. As they bid me goodbye they said I would be taking another trip soon—a much longer one.

They sensed the curiosity in my mind, though it was, I am certain, evident on my face, and one of them simply pointed at the horizon where a light of some sort was beginning to show. Not catching the meaning, I waved as they re-entered the craft, which hovered over my car for a few moments, then sped away to the west.

I turned the key and tried to start the engine, but it would not turn over. I thought my battery must be dead, or a terminal unhooked. Throwing back the hood, I checked the terminals, but they were tight. I unscrewed one of the filler caps, and steam gushed from the opening. I felt the battery and discovered it was very warm.

I got back into the car and tried the starter again. It responded and the engine started. Then I figured the proximity of the craft had affected the electrical system, as I remembered some of their previous statements. I made a mental note to ask my friends about it the next time I saw them.

I pulled out of the field and hit the main road. As I entered the highway I again noticed the light at the horizon, which, because of my different position, had turned into a bright moon.

It floated in a sea of clouds, and I remembered the trip they had mentioned. As they pointed toward the light at the horizon, did they mean to indicate that . . . ?

It was almost too exciting to believe.

CHAPTER 14

A CURIOUS ROUND TABLE

I stayed up very late on the Fourth, waiting for the promised phone call. It did not come, so I went to bed about 2:00 A.M. and almost instantly fell asleep. The next morning I awoke disappointed and wondered if I had been too sleepy to hear the ringing.

Feeling the plans had been changed I went to bed the next night rather early, but was awakened by the jangling of the phone.

"Howard, is it possible for you to make it tonight to Field Location No. 1? There is a possibility you can take some good pictures (though I sensed he had something more important in mind)."

"Give me a few minutes," I replied eagerly.

"Just a minute," he said. "Hello! Hello!"

"I'm still here."

"Howard, listen to this carefully. Wear dark trousers and shirt. Don't bring any flashlight. Don't drive up to the location, but walk."

"I understand," I told him.

"O.K., I'll meet you there. I'm calling from the Fountain Motel."

At that time I did not know the location of the motel, which I later found was on Route 22 near Clinton, and which figured in later contacts.

I dressed in dark clothing as quickly as possible, understanding why he had made the unusual requests. Neighbors, hearing a trace of my story, were becoming more curious about my nocturnal activities. One of them had stayed up all night, observing me from a darkened window, I heard later.

They didn't believe what they had heard about my meetings; instead they suspected I was up to some mischief. Their curiousity and mistrust made my work more difficult; and the resulting caution and secrecy I was forced to use gave my actions even more of an aura of mystery and intrigue. My position was becoming more and more uncomfortable.

I felt somewhat like a thief as I slipped into the pitch black night with Tassy, eager to go on another adventure, at my heels. She was a smart dog and knew she must follow me noiselessly as I climbed the steep hill to the rear of the house. Once in the open field, Tass took the lead as I struggled to follow her through the thick underbrush. We came to the wire fence which separated the steep hill from the contact area and crawled through an opening.

Once through the fence, I stopped a moment and waited. I couldn't see much of anything in the blackness, but could note that Tass ran ahead to the west end of the field, stopped and looked back to see if I was following, then sat down and waited for me to catch up.

Then a ball of light appeared over the hills to the west, moving in our direction at a very slow rate of speed. I took a picture of it. Tass looked in the direction of the oncoming craft, wagging her tail.

When it hovered over the crest of the hills, I snapped another picture. It came in close and moved slowly as if trying to help me obtain clear pictures. When it was almost overhead, I could see its undercarriage and snapped a third picture with a prayer that it would come out.

The craft landed, as Tass looked in its direction, whining. When the pulsation had subsided it was almost invisible to sight. Anyone coming into the field at that time would have sworn there was nothing there, least of all a space craft.

An opening appeared and a bluish, white light spread a path across the field and cast our shadows to the rear of us. A man's frame appeared in the doorway and beckoned to me.

I yelled, "Come back here, Tass!" as the dog ran yelping toward the ship. I was afraid the radiation from the ship would harm her. But she stopped short of the entrance, and by that time the man had stepped outside, had knelt down, and was petting her. As I walked toward them I could see that Tass was wagging her tail vigorously and whimpering with joy. Then I recognized our old friend who had made over her at the previous meeting.

The blond, Aryan-type man looked up. "Hello, Howard. Glad to see you again. Come inside."

I followed him and he turned to the dog and admonished, "You be a good dog and stay right there!"

Tass acted as if she understood perfectly, sat down and watched us quizzically.

We stepped into a large circular room. In the center of it was an ample-sized round table made of translucent material. Under the table top pulsating lights of many colors moved. The dowel-shaped stem supporting the table was set in what appeared to be a huge magnifying lens, itself set into the floor. Approximately one third of the circular room was devoted to instrument panels containing many colored lights blinking on and off. In front of the control board was a frame containing what I guessed was some sort of viewing screen.

A man at the control board was listening intently to what I supposed was a message coming from somewhere on the instrument panel. I couldn't understand the words, for the language was unfamiliar. Another man next to the operator was dividing his attention between the incoming message and our entrance into the ship. The two dark-haired men smiled and waved at us, then went back to their business—as I looked around me in amazement.

The blond man waved his hand over a section of the table and two of the lights stopped pulsating and two chairs came out of the floor. "Have a seat," he offered.

I asked him if the voice I heard coming in over their "radio" was from another planet. No, he explained, it was coming from another part of my own planet. They were contacting people all over the world and were in constant communication with their various points of contact. He said they could pick up pictures from any of the contact points on their screen, which looked something like a television screen, though perfectly square about two feet wide.

At that moment the man at the controls looked at me, then at the screen, indicating I should watch it. A picture appeared and the voices changed from the foreign tongue to English. I saw a tall, blond man in a space suit talking to a man with a cowboy hat and dressed in a business suit. The blond man turned, looked right at me, smiled and waved.

I was nonplussed.

How was it conceivable that the man on the screen could see us or know of our presence or know he was "beamed in" or being observed? The surprise was overwhelming, and no one explained it at the moment.

The man in the business suit appeared to be confused and uncertain as he kept looking around the craft in which they were conversing. Later I was told the fellow was a rich oil man from Abilene, Texas, who had been contacted by one of their people. His reaction was similar, they said, to all of us experiencing our first contact with people from other planets.

The picture vanished, and as I opened my mouth to ask questions, one of the dark-haired men moved over and joined us. I noticed his hair was held in place by a headband similar to the ones our American Indians wore; as a matter of fact had he worn a feather in the headband he could have passed for an Indian.

"Fellows," I told them, "I'm speechless," then went right ahead rattling off questions. Most of my queries they answered directly; others they politely ignored or led the subject away from them. I had learned never to press them when they chose not to answer.

Some of the information they gave me was quite technical and I wished I had a tape recorder with me instead of trying to rely on my memory.

"It wouldn't work on board this craft," the blond man said, picking up my thoughts; "all you would hear on the tape would be a hiss."

As we talked, a yellowish light flashed on the panel, then blinked off and on. All of us looked toward it, and they said it indicated a car was approaching the area. The blond man went to the wall of the craft where we had entered, passed his hand over a pulsating blue light to the right, and again the invisible door opened out of nowhere. He beckoned and I got up quickly and walked to the opening.

He asked me to tell the dog to leave.

"Go home, Tass, home!" I said, and the dog, still sitting near the doorway, whimpered slightly, then walked off disappointedly.

Again he passed his hand over the blue light and the door disappeared. We sat down again and he said we must leave the ground. Then the lights dimmed to a deep blue, then to a deep purple. The lens in the floor under the table lighted up. I could see through it, and it looked as if we were almost sitting on top of a dark gray Pontiac sedan.

"We're parked on top of a car!" I exclaimed.

"No, you're at least a mile above it. The lens magnifies it like a telescope," he explained as we watched the activities below.

The car door opened and a man and a woman emerged. The dark-haired man seated with us passed his hand over another of the pulsating lights visible through the table top and the image appeared as if in broad daylight. I could see everything clearly: even the blades of grass on the ground were clearly discernible.

I let out a little cry of surprise. I recognized the two people below!

The blond man motioned to the man at the control panel,

who turned something and immediately I could hear the two voices as if the people were in the ship with us.

The visitors didn't appear to be much interested in the scene below, but seemed to enjoy watching my amazement. The picture faded out and again I looked at them, this time without words.

A few seconds later the lens picked up their car again, as it was speeding along the highway. Again the picture disappeared and the room lighted up.

In what marvelous way were these people able to project color, light, sound? Surely it was a stupendous scientific feat. They seemed to take all of it for granted, as if the "miracles" were everyday accomplishments of their equipment.

They remarked they were surprised we had not discovered some of these things ourselves. We had the ability and resources, they said, but lacked the know-how in applying the natural, physical laws of science.

I was worried about my inability to fully comprehend all they were telling me. I had little or no education in the sciences. They told me not to worry, because many of their contacts were trained scientists who could understand their technology and now could even duplicate some of their instruments.

"It's fairly easy to find contacts like these, Howard. But people like yourself are a different matter. Remember the story of Moses in your Bible? He was gifted with the knowledge of how to perform miracles, but apparently he wasn't a good talker. That's why he had to have Aaron as a sort of publicity man.

"While scientists can perform certain tasks, few of them would be able to bring the spiritual message to people, as you will be doing."

They specifically mentioned a man from Yucca Valley, Calif., who they said was conducting experiments in some fields that were far advanced over our present-day scientific knowledge. I would meet him in the near future, they promised.

Would they give me his name?

"That's unnecessary. We'd rather let you have the fun of recognizing him, and it will be easy. Look for a fellow about six feet tall, well-built, blond, young-looking for his age, with a charming personality, and a great amount of understanding."

The man would be lecturing in New York soon, and they gave me the date.

"Just to make it a pushover for you to find him, take this down . . ." and he gave me the name of the hotel, the room and telephone number. I was to telephone him upon his arrival in New York and arrange to have a private talk with him. I was told that we might work together, along with some other people who had been contacted.

But again I was admonished:

"Remember, try not to let the story out until the summer of 1957, because there is still much training ahead. There are just as many people who do not believe and are opposed to the very thing that you know is true, and who will work diligently against you and your fellow contactees. Without proper instructions and training in the proper presentation of this information, you will be at a loss to handle the situations which will come up."

But nevertheless, they added they would assist me whenever it was possible.

"You know, Howard, a lot of our people are among you, mingling with you, observing and helping where they can. They are in all walks of life—working in factories, offices, banks. Some of them hold responsible positions in communities, in government. Some of them may be cleaning women, or even garbage collectors. But when you meet them you will know them!"

CHAPTER 15

GIFT OF AWARENESS

Our conversation terminated as the two men stood up. The blond man passed his hand over the lights, and the chairs disappeared into the floor. Putting his hand on my shoulder, he indicated I was to leave. The man at the instrument panel waved to me as the other man who had been seated with us walked toward the wall and again passed his hand over the blue light. The opening reappeared. The blond man stepped out and I followed.

Tass had returned and was waiting for me. As I shook hands with my blond friend, he said, "May you go in peace in the light of our Infinite Father."

A glorious uplifting emotion permeated my whole being, and again I felt humble and grateful for the privilege of being a small part of such a movement toward universal peace and understanding.

I looked forward to more learning. And my space friends had told me that with understanding and development, certain gifts would come to me. The gifts would be natural talents, which they had developed centuries ago on their planets, such as telepathic communication, control of matter, space travel, and a quite amazing one which I was soon to discover.

One night in my sign shop I teleported myself!

I was working late in the shop. Many nights I worked into the early morning hours because the increased number of contacts had slowed down my production, while at the same time the increase in family responsibilities demanded more financial resources.

Late at night in the shop I had peace of mind. It was a quiet place with no one to interrupt my aloneness with thoughts and problems.

I remember that I was working on a large sign containing letters about 12 inches high. The radio, tuned to an all-night record show, was playing softly. I was finishing the word, "SWIMMING," and was half-way through painting on the letter "G" at the end of the word when my thoughts wandered to Field Location No. 1, about nine miles away, where so many wonderful things had happened to me.

For a few moments I didn't realize my thoughts were

becoming strangely vivid. Then I realized I could see the field in its every detail, as the fluorescent light over my easel seemed gradually to dim and parts of the room faded out and finally disappeared. The light went from white to blue, to dark blue, and then everything blacked out.

The next thing I knew I was actually at Field Location No. 1, standing in the open field, with the cool, dewy grass beneath my feet. Time seemed to hang in suspension.

I was confused. What was I doing here? I tried to recover my bearings. Perhaps I had been called out on a contact and had left the house and come here; but how did I get here? Where was my car? Did I walk from the house? I walked around the field, hoping the fresh air would clear my mind up.

Then I remembered the sign, and the half-finished letter "G." The more I concentrated, the more I remembered, and finally I fully recollected.

Having concentrated on this location, somehow I had just appeared here—teleported a distance of nine miles!

I remembered that the space people told me we live in a three-dimensional world, an illusion we perceive with our physical eyes. They said the physical body is but a part of an illusion; that we really do not exist at all as we think we do.

The space people teach that the body is but a three-dimensional vibratory reflection or expression in the mind which the soul uses for a brief period on this planet. Most Earth people are prisoners of their bodies, while the space people control their bodies and other matter by use of the mind. They know and understand this to the degree that they can teleport themselves.

Man on this planet has lost most of his innate abilities he once possessed by leaning on crutches, such as telephones instead of telepathy; automobiles and other vehicles instead of teleportation, and so on. He has turned toward materialistic, unnatural energies, instead of utilizing natural laws given to him by our Infinite Father to be used freely.

Yet people on this planet consider themselves highly civilized. Actually they have been going backward for centuries. It is true we have more gadgets, conveniences, TV sets, radios, telephones, electronic machines, computers—and atomic bombs. But all of these things man creates only limit him. He is becoming a slave of his own inventions. How true was the expression of some wag when he twisted an old saying and came up with, "Invention is the mother of necessity"! There is little need for complicated gadgets which we call conveniences. The mind is more versatile and powerful than any gadget ever created, if only we would use it.

Our space friends cannot understand why we make things so complicated, doing things the hard way while calling it the easy way. They take primary energy from electrons present in our atmosphere and channel the force which surrounds each electron into all types of simpler conveniences which cost nothing to operate, utilizing, so to speak, a free energy. But we generate electrons, first, through brute force, then run the electrons along insulated wires where they again, through brute force, operate our complicated gadgets.

The space people do not destroy the electrons; they merely take the power which surrounds each of them and use it constructively; then the released electrons go on their way, recharge and become usable again. Our method of handling electrons destroys them, for they are dissipated as heat in electric lamp bulbs, X-ray machines, radios and other electrical equipment. Some of these operations throw off the toxic elements and damaging rays, which reminds us again of the principles of natural law. When we go against natural law, the results react against us.

The space people had promised to show me instruments with which they can teleport any kind of matter, including human bodies, any distance, providing there is a transmitting and a receiving set at the points of departure and arrival. They could not understand, they said, why we had not accomplished this feat ourselves, since we were so close to it with our transmission of light energy in the form of images. In our now quite ordinary television sets we are able to convert light from three-dimensional bodies into electrical energy, then back into light energy and images. The secret of going much farther and transmitting matter itself is within our very grasp; in fact some scientists have actually done it secretly.

Everything one creates must originate as thought, which consists of vibratory reflections under control of the mind or intellect. Whatever man thinks, he can do. People who have the ability to teleport themselves also have the ability to use this potential to the highest degree. Telepathy is not, according to the space people, the mere sending and receiving of brain waves. What happens is a realization on the part of two individuals who are communicating that they are not limited by their physical bodies and they are not in a time field. They are in a timeless zone and are aware of all. Those who use it now and then are in both the physical third-dimensional and the fourth-dimensional reflection, without being able to break completely with the third. They experience an awareness of something other than physical, yet are still conscious of the third dimension.

In ages past certain glands under the brain were developed by people who at that time on this planet could use the glands frequently as an aid to telepathy and other natural talents. These glands are now just degenerated and underdeveloped, just as man's muscles become lax in absence of proper exercise.

The brain is an instrument of the mind. We communicate with each other by telephone in a round-about way. The soul, expressing in the body, communicates its desires to the mind. The mind sends out thoughts through the brain. The brain converts the electrical energy into sound energy via our vocal chords. The sound is picked up at the receiving end of our telephone by a system of magnets, which changes the sound energy back into electrical energy; the electrical energy travels along wires to the other end of the circuit and is picked up as electrical energy and changed back to sound energy. The latter we perceive with our ears. Our ears transmit the sound energy to electrical energy through the brain, a physical, complex blob of billions of atoms spinning around each other, just like our solar systems. The electrical energy is then converted into thought patterns which translate the meaning to the mind. The mind, the communicator between soul and body through the instrumentation of the brain, is the recorder of our very existence.

Teleportation is relaxed photographic perception in three-dimensional detail, involving light refractions, shadows, life sound, objects, smell, and in general, using all our five senses, plus the sixth one which involves utilizing the gland beneath the brain. In other words, you must perfectly visualize or image the scene to which you wish to be teleported; then you mentally become a part of the scene.

Thus I tried to explain to myself my sudden and surprising change of location. As I continued walking and again thought about the shop and what I was doing there before coming to the field, I again experienced a peculiar sensation.

First a bluish-green haze enveloped me and the more I thought of the shop the clearer an image of it became. The field and trees became indistinct and hazy. The ground below me disappeared. I was not conscious of my feet or my body. I felt as if I had one huge eye, and that I was the eye, aware of the magnitude of space—unlimited, unending.

Then the next thing I knew I was sitting on the chair at my easel in the same position as when I had "left," with the peculiar feeling that I had just dropped my brush in the middle of the work and was reaching down to pick it up. I bent down to get it and found the brush was dry and hard. The unfinished "G" I had partly painted was dry!

Then it was not my imagination. I had really been gone!

Knowing the paint always required 15 minutes to dry, I knew I had been gone at least that long. Could it have been a dream? Had I fallen asleep? Had I done so, I would have fallen forward, or at least rested my head on the sign, which would have smeared the paint and ruined the sign. I would have got paint on my face and arms. Yet everything was intact!

But I was never certain of just what had really happened until the matter was later clarified and explained by my friends.

CHAPTER 16

TRIP IN A VENUSIAN SCOUT

Around the first of September, 1956, I again went to Field Location No. 1. The many contacts had somewhat dulled my excitement until the awe-inspiring meeting the week before. Now as I saw the ship coming in, I hoped I might have another surprise only half so wonderful; and in the back of my mind was always that small hope I would at last get the promised ride into space.

The bright lighting faded and I could see the ship in detail. It was much larger than any of the others I had seen, and the actions of the two men who got out of it convinced me that something "WAS up."

They invited me on board, and I stepped into a large circular room. Again the light seemed to come from the walls, and the interior was similar to the one I had visited a few weeks previous, excepting this one was larger. The size was first evident to me when I noted the many chairs around the center table, contrasted to the three visible during the former visit.

The table was also somewhat different. The most striking difference was a large coil-like apparatus, apparently made of gold, which rested on the center of the table. I assumed it had something to do with the lens beneath the table, though when I entered the ship the latter wasn't visible. Apparently they made it transparent and operable by "turning it on" in some fashion.

Again three men made up the crew, yet I could see shadowy outlines of other figures beyond the lighted translucent walls in the corridor surrounding the circular room. I was not invited beyond the one room. Again one of the men sat at an instrument panel, another man stood near him, and the third man sat with me at the table.

None of the men had said much, and I had learned it was a better idea not to ask questions—to wait until they were ready to tell me what I should know.

"We're going to take you for a trip beyond the earth's atmosphere," the man at the table remarked abruptly. Then he added, "We won't have any peace until we do—you've been after us constantly telepathically."

The very serious look on his face changed, as one corner

of his mouth twitched as if he was going to smile, and again I knew he was engaging in the "dead-pan" humor some of the space people liked to effect.

The view screen lighted to a beautiful view of the heavens, with planets greatly magnified against a magnificent background of stars. I saw a fleeting glimpse of the moon.

"Was that the moon?" I asked him.

"Yes, we just passed it," he replied in a matter-of-fact manner.

For the first time I realized we were already out in space. I had felt no jerk when we took off and thought we hadn't yet started. He was amused at the disbelief in my face.

"Watch the screen," he told me, "and you'll see an interesting show in just a moment."

Then I could tell we were moving by looking at the screen, since the stars seemed to be traveling slowly, horizontally across the screen. I did not know how the screen was oriented in regard to direction and half expected to see the stars flashing past us. To this day I do not have a good knowledge of astronomy.

Suddenly I flinched. A huge rough-shaped object headed toward us!

Again the man laughed at my amazement and I assumed we were in no danger. The object suddenly veered away from the screen and disappeared outside its field. Then I noticed other smaller objects which also veered away.

"Are those meteors?"

"You're correct. They aren't quite as close as they appear, however, though our great speed makes it appear we're encountering a concentration of them, when actually they're far apart."

Nevertheless it was still nerve-wracking to watch them.

"Don't worry, Howard; should one move on a collision course with our ship the flux around our hull would repel it and it would veer off to the side."

Just then the man standing beckoned to me and I turned to my friend at the table, seeking permission to leave him. He nodded and I walked over to the other man. He pointed toward the wall where a porthole appeared in the same iris-like way their doors opened. Through the porthole I saw a hazy, whitish, fluorescent object which reminded me of a tennis ball hanging in a dark sky.

I asked if it were Venus or the earth.

He smiled and waved his hand over a section of the instrument panel.

Again I gasped in amazement.

A few years ago I took my family to see Cinerama, the new motion-picture process which employs a wide, deeply curved screen. When the picture began my family and I sat there watching Lowell Thomas giving a lecture from a small black-and-white picture. But suddenly the curtains opened onto the famed roller-coaster scene, and the hugeness and apparent reality of the giant picture suddenly bursting upon me, almost took my breath away. I grabbed onto my chair as the roller coaster made a dip, while my family laughed more at me than at the picture.

That is the best way I can describe my amazement at looking toward the center of the room and seeing it change into a three-dimensional picture; but it wasn't flat like a screen—instead it seemed we were actually flying a few feet above the surface of a beautiful planet. Immediately I saw it was not Earth, and my question had been answered: Venus!

The scene shifted rapidly. Sometimes we were about 10 feet above the scene below; at other times we were higher, possibly a hundred feet off the surface. I saw beautiful dome-shaped buildings, with tiers spiraling upward.

The planet was fantastically beautiful. I did not get the impression of cities; instead, I was reminded of beautiful suburban areas I have seen on our own planet, though, of course, wonderously different. The buildings were set in natural surroundings with large trees, which looked something like our redwoods, and gardens stretching in every direction. Then I saw forests, streams, large bodies of water. People, dressed in soft pastel colors moved about. I also saw four-legged animals which were unfamiliar to me.

Vehicles moved on the surface, apparently without wheels, for they seemed to float slightly above the ground.

The operator of the marvelous picture show saw my interest in the surface vehicles.

"No, they don't move on wheels. We have very little use for wheels. In fact we didn't go through the wheel evolution which has, in reality, slowed down your civilization, instead of having been the great boon you have been told about in school. You would have done better to have omitted it—as we did!"

The picture swirled and grew blurred, then disappeared, and again I could see the center of the ship.

"The show's over," he laughed; "and don't ask me for a double feature!"

Almost before I knew it the door opened again on Field

Location No. 1. The entire trip could not have lasted much longer than half an hour, though I had forgotten to look at my watch when I entered the ship.

I walked the familiar route out of the field. Being back on earth again after seeing a little of what lay beyond was almost like going to prison.

CHAPTER 17

THE WITNESSES

My work gained in momentum and the contacts were even more frequent. Now I proposed to invite others to witness the craft and their occupants.

Late in the summer of 1956 the visitors suggested a list of possible witnesses and indicated the time and location of a proposed landing.

The big night turned out to be a dark, sultry night in late August when I brought a small group up to Field Location No. 1.

A sense of excitement pervaded the air. The witnesses consisted mainly of people from the surrounding community—but one was a physicist from a large eastern university. My own experiences had taught me to understand the emotions and feelings of people who were catching first glimpses of the visitors, so I watched over them closely and reiterated that whatever they saw, not, under any circumstance, to make a wild dash toward an unknown object or person—but to wait and observe. I knew that overhead a craft, dark, and out of sight, hovered—waiting.

Then, on the ground, near the edge of the woods, pulsating lights appeared. We halted. I received the mental suggestion that it was all right to approach closer, and we moved toward the lights. Now we could see small disc-shaped objects of various sizes ranging from eight inches to eight to 10 feet in diameter, with differently colored lights pulsating. I knew these were observation discs similar to the one I had seen in April and that they were busy recording the thoughts, emotions and intents of the witnesses. I explained what the discs were, without elaborating on their recording mechanism.

I remained emphatic that they were to approach no closer than 15 to 20 feet, though I hesitated to alarm them by telling them of the magnetic field.

A few nights later I was told to invite the same witnesses again, since the visitors felt it was safe to allow them to have an actual look at themselves.

The visitors, landed about a quarter of a mile to the rear of the house in a secluded wooded area. They walked down to where the witnesses were waiting, hurdling a fence into the apple orchard, where I advanced and spoke to them in full view of the witnesses. These men from another planet were very tall, close to seven feet; and I defy any Earthman to equal the physical abilities they displayed to the amazement of the spectators.

One of their feats was leaping and at the same time giving the appearance of gliding over a fence five feet high and covering a distance of 20 or more feet in an abnormally short time.

Once when headlights of an oncoming car on the road passing the property flashed across the orchard, the visitors jumped and ran about, as if trying to avoid the glare of the headlights. I learned later that the glare of our artificial electric lights, as well as our harsh daylight, is disconcerting and annoying, and in some cases, painful, to some of our visitors. The hazy, cloud-like atmosphere shell which once surrounded this planet has now gone, and as a result we receive the direct glare of sunlight without the protection of a natural filter.

I walked toward one of the visitors, shook hands with him and exchanged a few words in English. I then turned and walked with the man, who was about a head taller than I, toward the spectators. He halted at a distance of about 10 feet from them. At the same time the witnesses were observing two other men and a girl.

I believe I can better describe one of the witnesses' reactions by presenting a transcript of part of a radio program. It was to be one program in a series of unusual programs which would bring thousands of people to see me and hear my story first-hand.

I am referring to the "Long John Party Line" show on Station WOR, New York City, presided over by John Nebel, known affectionately to millions of listeners as simply, "Long John."

Nobody remembers exactly when Long John first began talking about flying saucers, but when he did he found his listeners preferred that he stop playing records altogether and spend the night, from 1:00 A.M. until daylight, interviewing interesting and controversial people.

Long John would shockingly cry, "I don't buy that," if I stated my belief that the space people themselves led him into pursuing the unusual format, thereby acquainting millions with startling and unfamiliar New Age subjects, which they subsequently took to heart: for he is still a skeptic regarding my story and other experiences he had presented.

But I have a great debt to repay to Long John. In those early days he let me tell my story to his huge audience. And constantly reminding the listeners, "I can't buy that, but—" he would just as often leap to my side when other members of panel shows would try to confuse me by cross-examination.

I know the testimony of one's own father is not considered as acceptable as that given by a non-relative—but I take the liberty of doing so, in loving memory of Dad, who passed away in the early fall of 1957.

Here is part of the interview:

LONG JOHN (addressing my father): Would you say these were normal-sized people?

DAD: Oh, no. I would say one was about six-feet-two or three, and the other was about six feet. . . .

LJ: Were you close enough to observe their features?

DAD: No, that I wasn't.

LJ: Did you notice what they wore?

DAD: To a certain extent, yes. As far as I could see, they wore something similar to ski suits, tight at the wrists and ankles, and the rest of it seemed to be . . . well . . . I just don't know how to explain it.

LJ: I think you are doing an excellent job. Don't worry about the choice of words . . . everybody has a different view of things, like the three blind men who examined the elephant. You know what I mean . . . each person comes out with a different story, and I don't want to be unkind, Mr. Menger. For instance, that green sport shirt you have on: somebody else might say, "I didn't notice the green sport shirt, but I liked the striped, double-breasted suit he had on." Everybody notices something outstanding about an individual. I think you will agree. What kind of night was it?

DAD: It was a dark night . . . but these people seemed to have a glow to them. That is how we discovered they were coming toward us—by the glow. When these people left us, the grass there (and I know positively because I had cut some of it) was three feet high. And they went through that grass like it was a nice concrete walk, with no exertion at all.

LJ: But afterward . . . the next day when you examined this stretch of grass . . . was it matted down?

DAD: I'm sorry, I paid no attention to that.

LJ: You have seen the ships, too, haven't you, Dad?

DAD: Oh, yes, I have seen them in the air and in the daytime I saw them, and at first I was very skeptical.

LJ: You are not any more, are you?
DAD: No, not now. . . .

And I would like to think that Dad left this earth knowing at least a portion of the truth.

CHAPTER 18

THE POLICE

One night as we sat at the table having a last cup of coffee and snack before retiring, a knock came at the door. We opened it and four men entered.

Three of them said they were detectives from a nearby state police barracks; the fourth was a newspaper reporter. They assured us they were on "unofficial business," that it was merely a social call. They had heard some of the stories already circulating, were intensely curious, asked to hear of my experiences first hand.

Barring the fact they called at a late hour and were obvious in their intents, I patiently went through the story of my contacts, the sightings and the witnesses.

They asked to see the pictures and I had to tell them they were not available at the moment. That made them even more skeptical. The pictures had been loaned to a friend, so I offered to go and get them. After telephoning my friend and learning the family was still up, I got into the car and left. On the way to Wood Glen, where my friend lived, I heard a voice speaking my name over the car radio:

"Howard, show your friends the pictures and after that we suggest you show them Location No. 2, for there will be a ship in the area and a three-foot disc on the ground approximately 100 yards from the entrance to the field. Three men will be there to greet you. We suggest your friends bring no flashlights, nor weapons of any kind."

The voice was gone and the music, interrupted by the message, came on again in full volume. At that time I did not recognize the voice; later I would meet its owner in California.

When I returned, I showed the men the pictures, and although they expressed much interest, I could sense their skepticism, though they tried to hold it politely in reserve.

I asked them if they would like to see "something" at one of the contact points, and had never seen anyone look so eager as they did. I asked that they leave their guns on the table or in their cars, and that they not bring flashlights. They assured me they were not armed.

We went up to Location No. 2, and just as we entered the field, I saw a pulsating glow ahead of us, and I made out the faint outline of a craft above the trees. We stopped and got out of the car. Two of the men admitted seeing the glow, but not the outline of the ship. I asked them to wait by the car while I went ahead to see if it were all right for them to come ahead. I greeted one of the space visitors who was waiting for me. Farther down in the field I could see two others, and near them an observation disc on the ground.

The man I greeted was dressed in Earth clothing; Leather jacket and trousers; and his hair was trimmed in our style.

Suddenly the disc on the ground changed from a bluish-white to a reddish-orange color and I knew what it indicated. I wondered if it were I who created the negative reaction. The man answered, "No, it is the gentleman by the car, and it is because of his ignorance and fear of the unknown. We are sorry we cannot invite your friends to come any closer, for there is a weapon present, and one of the men would attempt to use it without hesitation!"

I was discouraged. The men had lied when they said they had no weapons. I thanked the visitors and walked back to the waiting men. I told them the disc was there, but advised them not to go any closer. They complied and smiled knowingly at each other, and I could tell they didn't believe me.

I filled my pipe, but when I reached into my pocket to get my lighter, discovered I had left it at home. Remembering a spare in the glove compartment, I reached in to get it, and, as I did so, picked up the thoughts of one of the men: he thought the glow they had seen had been caused by my lighter and that I had pretended to have left my lighter in the car. They thought I had perpetrated a hoax.

Many of the witnesses were interrogated by various civilian investigators who showed a desire to prove or disprove my story. The best investigation was carried out by Jules St. Germain, an attorney from Lynbrook, Long Island. His intelligent and fair approach to the entire matter was respected by the witnesses and myself. Even though hundreds of people besieged our house, St. Germain managed to take the witnesses aside and tape record their versions of the story.

The best tape, in my opinion, was the account given by the physicist. The witness patiently talked into St. Germain's microphone for two solid hours, and gave a completely unbiased account.

One night as St. Germain, Lee Munsick and another man were leaving my place, they said they were going up to

Location No. 1 before they left for home, hoping they would see something. I smiled, for I knew that my space friends were making themselves and their craft obvious for as many witnesses as possible during this particular phase of my story and felt they might actually see something which would lessen their skepticism.

"You will be back," I told them as they left.

Thirty minutes later we heard them again at the door. Two of them looked perplexed, the face of the other was white. Something had happened up there, but I found it impossible to draw anything from them.

It is difficult for people to absorb what they do not understand. It is often easier simply not to believe than to accept. Many strange things, involving actual contacts with beings from other worlds, have happened; often these have seemed inexplicable and overly mysterious to some people. They have often been definitely alien to our concepts.

I have made many errors in presenting this story and in accomplishing some of my missions. I have made marginal errors, over an underestimation of people, situations and experiments. Many times I have been bewildered in ascertaining the correct way to handle circumstances, and to many of my friends I have thus appeared confused. I was trying to do a gargantuan job with feeble tools. However, I have never given up hope or faith, and the pattern of my life is now assuming a more stable atmosphere, which, in the future, will be conducive to more constructive work.

Higher civilizations can actually destroy a lesser one without intent, by imposing their higher wisdom on those who cannot understand or absorb it. The space people realize this danger, and they are exercising great care in the manner of going about our education.

When the white man came to the Indians, and when missionaries went to the natives, they superimposed their concepts and methods on less-advanced civilizations. Many times the results were disastrous. The space people have perfected many machines, instruments to aid in better living, gifts of awareness, which would be of tremendous benefit to our humanity. But if they gave these to us abruptly, without preparing us gradually to use them, these gifts might bring disaster instead of blessings.

Too many people, for example, who, after receiving gifts of awareness and developing certain God-given powers, such as extrasensory perception, astral projection, teleportation, and so on, use these powers to further their own selfish desires—to gain control and influence over their brothers on this planet, instead of helping them to help themselves. I personally know several who have developed certain of these talents and have then used them for curious experimentation and for impressing people with their "unusual powers." Such powers, unless used ONLY when necessary, and ONLY for helping others, will be taken away as quickly as they were given—or they will be gradually dissipated over a period of time with serious ill effects. I most urgently and humbly implore my brothers who are gifted with certain powers to use them sparingly and only in the name of our Infinite Father. In my own early experiences with these gifts I, too, made mistakes, but I am thankful that I have now realized the correct manner in which they should be used.

CHAPTER 19

THE PHANTOM CAR

Many investigators came to check my story and question my witnesses. When I learned that a certain man was to give me a "going over" I was certain of one thing: Although I would be gone over with a fine toothed comb, I would be investigated fairly. That man was John Otto, one of the best-known civilian researchers in the flying saucer field.

His initial skepticism, I can happily report, ended with sponsoring me at a lecture in his home town, Chicago, Ill. I think it best simply to quote Otto's introductory talk verbatim, for it describes his investigation better than ever I could—in his own inimitable way which has made him such a popular lecturer:

. . . Now we come to the part of the story many of you have asked about: just how I happen to be sponsoring Mr. Howard Menger. Let me start by explaining how I happened to meet him.

I had just finished trying to stabilize one of the fraudulent stories on the west coast. I had been requested to go there and check this thing which turned out to have been a prank originally—which someone had whipped up and amplified into a terrific fraud. In order to keep my nose clean in the field of research, I had to dig up the facts, though I did not like the job—for I do not like to look for untruths, but to look for the truth. However, one of the best science editors in the newspaper field had asked me to do it and I couldn't turn him down.

Shortly thereafter a telephone call came through from New York City: "Please, Mr. Otto, come out here and apply your investigative technique and your knowledge gained from nine years of research in the field, and EXPOSE ANOTHER ONE!" I did not want anything to do with it, and I thought, "This time I have HAD it, and I am ready to quit. If I have to run around the country and prove this man valid and that man a crook, I want no part of this subject."

Finally they said they had to have somebody come there and do something about it and they would pay my plane fare. So I thought it over; then a business trip worked in whereby I could pay some of my other expenses I would incur. So the short of it is that I went east to investigate one Howard Menger—and I got a jolt! A severe jolt!

When I got into New Jersey I spoke with Howard Menger and still could not believe it. He could not show me evidence I could consider conclusive, and if I am easier to, shall we say, show evidence to than a lot of people, I have been in this a long while, and I know some of the truths in it. A short while after this I had to make a broadcast on the Long John all-night radio show. On this show several members of an east coast investigative organization were already damning this man Menger without any investigation at all, merely on the grounds that the man had claimed he had met interplanetary visitors. It was awfully hard; it was a difficult situation. They asked me what I thought of Mr. Menger, and I had to tell them honestly that it was too short a period of time in which to accept or reject a story of that nature.

Two days later I was requested to try another contact attempt (such as Otto had successfully carried out over Chicago's WGN—H.M.) in a broadcast over WOR, a 50,000 watt station. At about four o'clock in the morning of January 10, 1957, we again tried to establish a communication contact that changed the whole script suddenly. I decided at this moment, that if the boys wanted to come through and help, then they could help this way: let's have an endorsement of Howard Menger! Endorse him—or reject him, whichever the case might be.

And rather astoundingly, I received an endorsement of Howard Menger! Not conclusive, but enough to turn me overnight into a supporter of this man, who, a few moments before, I thought was a fraud! Everybody at the station received a shock. The next night Long John received a telegram from me stating, "I may have evidence that may validate Howard Menger."

Then I went back on the air, but was not in accord with submitting the evidence—I didn't want to. I thought it best to keep this evidence back at the time, but I was forced into submitting it and, with Howard Menger's permission, we submitted that evidence over the air—but not all of it—just a minute part of it, the kind of thing we as Earth people must always look for . . . a little bit of truth, a little bit of evidence that we need to go along with this thing.

Now I want to ask you all, right now, not to plunge into acceptance of everything in the flying saucer subject. Examine it carefully. Evaluate it just as we have done; and then, when each and everyone of you are ready, believe me,

YOU WILL HAVE YOUR OWN PROOF—ALL THE PROOF YOU WANT!

The end of this section about my witnesses would hardly be complete without my relating one rather humorous incident.

Our group, of about 40 people, met every Thursday night at the home of two of the members, a charming young couple of Pluckemin, N.J., who graciously donated their recreation room for our meetings. At these gatherings I usually talked about my experiences, but my main purpose was to present spiritual concepts as taught by the space people. My job in conjunction with the space brothers was to awaken within each member a desire for learning more about the universe and its true meaning; our purposes on the earth; where we came from, and where we are going. While it was an exhilarating mental pursuit, some minds were able to absorb it, but some were not. It was religious, but only in the sense that we tried to present the theme that God is omnipresent, omnipotent and omniscient. Out of this small test group the space brothers and I hoped to produce teachers who would teach others who, in turn, would teach still others. The meetings usually began about 8:00 P.M. and lasted until midnight. After the lesson, we would have a coffee break and then a general social get-together before leaving.

One night in the spring of 1957, during a coffee break in the discussions, I mentally broke away from the group to relax. My thoughts took me back to the old 1950 light green station wagon I had traded in a few days before in Philadelphia for a new 1957 station wagon. I sentimentally thought of the "old jalopy" and the many wonderful experiences I had while owning it. In my mind's eye I drove it along on a blacktop road, picturing many things in vivid detail. Then I left the reverie, returned mentally to the group, and becoming aware of the discussion, joined in, without giving another thought to my vivid mental experience.

We left the house about 12:30 A.M.

At the next meeting the telephone rang about midnight and the host answered it. I was surprised to learn the call was for me from the police station in Bedminister Township, a few miles from Pluckemin. I picked up the phone.

"Are you Howard Menger?"

"Yes."

"We have a summons down here for you. Would you

please come down and pick it up? Sgt. Cramer claims you were speeding and went through a red light in his district about 11:40 P.M. on (he named the date of the last meeting)."

Checking the date and time with the group, I said, "It couldn't have been me, because I was here at that time and there were at least 20 people here with me. Besides, I do not have a 1950 station wagon, sir; I have a 1957 Plymouth station wagon, and incidentally, it could not have left the premises because it was blocked in by other cars, and I had the keys in my pocket."

The voice insisted I come down and pick up the summons and appear in court to answer charges—else pay the $15.00 fine.

I did not go that night, but after thinking about it, I realized what had happened. I had been thinking about my old station wagon and had been driving it mentally the night of the last meeting—at the exact time mentioned by Sgt. Cramer! Could it have been possible that my thoughts had manifested into an actual projection? They finally sent the summons through Police Chief Kice at High Bridge, who delivered it personally to my home.

I decided to appear in court and took along seven witnesses. The charges were presented; then several witnesses testified, and finally I was called to the stand. I pleaded not guilty to the charges. I said I was not there at the time and had witnesses to prove it. Furthermore, I did not have occasion to drive through that section when going to and from my friend's house.

Sgt. Cramer's testimony went something like this: He saw a light green station wagon with license plates WR E79 speed past him. When the car reached the red light at the intersection it went right through without stopping. He said he pursued the car to the red light where it "disappeared."

His using the word, "disappeared," intrigued me. The roads stretch out straight for long distances through the country in each direction from the intersection, so it was likely that tail lights from such a speeding car would have still been visible. He was asked if he saw a driver, but he said he did not, and reiterated the station wagon had just disappeared at the intersection.

When the judge heard that, he remarked. "Well, what do we have around here, a phantom car!" A tenseness came over the entire room. I realized Sgt. Cramer was not lying—he had seen my old car, and I felt sorry for him.

The Judge said, "I feel like either putting a man in jail for perjury or breaking a sergeant!"

I was finally recalled to the stand and still affirmed I was not guilty. Finally, the testimony of the witnesses and fact I no longer owned the station wagon gave the judge no other alternative but to arrive at the decision of "Not Guilty."

Incidentally, we had checked with the auto agency in Philadelphia where I had traded in my old station wagon and learned it was still in the shop being repaired for resale.

The judge aptly summed up the case with:

"This is the strangest case I have heard in all my years on the bench!"

CHAPTER 20

THE SONG FROM SATURN

Although by the fall of 1956 I had become accustomed to encountering the unusual everywhere I turned, an old cabin in the woods, secluded and obscure, was the center of one of my most bizarre experiences.

As I stood there, hesitating to walk up to it, I knew it was the only building for miles around. And I knew something had drawn me there, because a few minutes previously, I had suddenly lost control of my car, and realized it was being driven for me by some higher intelligence.

This new experience was indeed unique, and it opened to me a whole new world of creative expression—one I had no talent for, or thought I hadn't. Considering the work of the space people, however, I should not have considered it unique, for I had found that continual contact with my friends from outer worlds have seemed to trigger in my consciousness many hidden talents and gifts of which I had been completely unaware.

The car, whatever was controlling it, had let me off the main road, and I had ridden for miles through the countryside on little-used roads, not knowing just where I was. When the car came to a stop I looked off through the woods, and there to my left was the old cabin, dilapidated and apparently unused for years.

Whomever or whatever was in the cabin, I knew I had been brought there and that I must enter the old dwelling.

I walked nearer. Very low at first, and then louder, the strains of the most inspiring, soul-tingling music ever to fall upon my ears emanated from the cabin. I paused outside the door, stood there entranced, letting the music flow through me. I seemed to be absorbing the music into my entire body, and my heart almost beat in time to its pulsating rhythm.

The music seemed to soothe and excite me at the same time. Yet I found a bewilderment developing within me, something that distrubed me deeply. What was it? Then I knew the reason. Somehow the music was familiar, but only vaguely so. Had I heard it somewhere—on the radio, on a television show, in the movies? Certainly not this kind of music? Once I had stood suddenly in a great auditorium when an entire audience had risen to its feet as a great conductor, now passed on, waved his baton and brought wild crescendos thundering from the brass and percussion sections of the orchestra. Under his superb direction we heard sounds and emotions never before wrung from the heart of musicians. And some of us had wept.

But this music was not a wild crescendo of sounds, nor a heady emotionalism coaxed from strings. It was a simple melody, apparently played on one instrument.

I pushed timidly on the door, partly ajar. At my touch it swung open, and there, sitting at a piano-like instrument in the middle of the main room was a man, playing the unusual melody.

Little about the man appeared strange. He wore a rough woolen shirt, just as any camper might wear, and trousers tucked into heavy boots. He was ordinary enough, excepting, perhaps, for his long brown hair, which curled under and fell to his shoulders, reminding me of pictures of page boys. His skin was smooth and white, and his eyes, which I first saw when he looked at me and smiled, were hazel. His entire demeanor was one of serenity mixed with good humor.

My eyes went from the man to the surroundings. The floor of the rustic cabin was wood, partly covered by a rug. At one end was a huge fireplace, and at the other end I saw a series of instruments which I realized certainly didn't originate on Earth. From the instruments my eyes went to the wall on which I saw a clock, but a most unusual timepiece. Instead of numbers I saw 12 fluorescent spheres; nor were there hands on the clock: instead I noticed a very bright light on the face of the clock where the hour hand should be. I glanced at my watch, which said 10 minutes to four, then back at the clock and noted the bright light appeared very near the fourth sphere. Where the minute hand should be I saw another light, of lesser intensity.

On the floor were more instruments, one of them a box-like instrument with a view screen, something like a portable television set. Another instrument was shaped similar to a console set, with a coil-like aerial revolving on top of it.

Just then two blonde-headed men stepped from another room, as one of them greeted me, "Hi, Howard, we have been expecting you." Then he introduced the other blonde man and told me both of them were from Venus.

"And our own private 'Liberace' here (as the pianist scowled at them, then grinned) is from Saturn."

The Saturnian broke off playing with a good-natured, theatrical flourish on the keys, rose and extended his hand. I complimented him on his playing, and told him the music sounded familiar. Then I halted in the middle of a sentence, for I remembered where I had heard the melody. Why, it was the little tune I had so often hummed, and even tried to pick out, unsuccessfully, on the piano. I had never given the melody much thought.

"Sit down and play it, the melody," he offered, gesturing toward the instrument, but I stammered the only thing I could play was a phonograph. He put his hand on my shoulder and spoke reassuringly.

"*This* music you can play, Howard," and he guided me gently toward the chair in front of the "piano." I sat, protestingly.

"From this time on you will be able to play a piano whenever you are moved to do so, and not only this tune, but any melody you wish."

I looked down at the keyboard. It was entirely different from a conventional piano keyboard. This one was much longer and contained many more keys, which were narrower and had strange symbols on them which I did not understand. The entire instrument was much lower and closer to the floor.

I almost automatically reached down to touch the keys, suddenly knowing which to strike to correspond with the sounds of the melody running through my mind. Although I had never been able to play before, it all seemed natural and delightfully simple.

Although I could not master the subtle nuances I had heard in his playing, I believe I could have played it otherwise as perfectly as he, had my fingers not often struck two of the keys instead of the intended one, probably because of the narrowness.

As I played, the men looked at each other and nodded; then they applauded politely when I finished. I was thrilled and happy, because I knew I had suddenly found a beautiful, haunting melody that had always stirred my imagination, and found myself playing a musical instrument for the first time.

The Saturnian spoke. "You're wondering why we brought you here for a musical exercise, and it probably seems foolish to you. But it isn't. You're going to play this melody on the piano, Howard, and thousands of people of Earth will hear it."

I visited with them all that night, which we spent talking of many things which would be taking place in the months to follow. Early in the morning we parted, as one of the blonde men saw me to the door and jokingly dismissed me with, "Farewell, Maestro."

It was still dark, but I started up the car knowing I would be shown the direction back to the main roads. So I just drove wherever my subconscious indicated, and in a very little while recognized a familiar highway. Once or twice I have tried to find the wooded area again when I have been in that section, but without success.

I couldn't wait until I could find a friend's house where there was a piano, where I suddenly walked to the instrument, sat down and played a popular song.

As all my friends would exclaim later, this particular one remarked, "But, Howard, you didn't tell us you could play!"

I had been told that anyone hearing the music the Saturnian and my soul had taught me would get a feeling, or reach an awareness, which would act as a mental assist to release something from the subconscious. People hearing the theme would react in their conscious state with increased understanding and brotherly love toward one another.

Every musical note, they had explained, has its specific density and frequency which causes a sympathetic vibration, when created at the correct frequency and in certain combinations (witness a glass breaking in the presence of a high-pitched musical note). This, of course, is an oversimplified example of what happens to the subconscious when certain music is heard—but its principle is similar; in short, sound causes a corresponding effect in one's innermost mind.

For many months I played the music all over the country, as I noted it did have a noticeable effect upon those who heard it. I taped the music and sent it around to various study groups; but knowing that few people have access to tape recorders I finally was able to have a long-playing record of the music issued for the public.

The congenial president of Slate Enterprises, Newark, N.J., was so impressed by the music he suggested we produce the record, which we titled, very appropriately I thought, "Music From Another Planet," now available on Slate Record No. 211.

In time I realized that whenever I sat down at the piano and allowed my fingers to stray, other music came through.

This became as much a surprise to me as to my friends and relatives who had never heard me play before.*

This was but one of the gifts awakened in me by my brothers from outer space, to whom I am eternally grateful.

* For years I had tried to pick out melodies on the piano with little success. [The author has subsequently recorded this music in a long-playing record album, *Music From Another Planet,* distributed by Space Records, Box 2228, Clarksburg, West Virginia 26301.—Ed.]

CHAPTER 21

MARLA

In the excitement and confusion I had almost forgotten what my friends had told me of a certain man from Yucca Valley, Calif., whom they promised I would meet.

In October of 1956, however, I must have been led to go through a certain batch of my now huge collection of notes and data; for there on a scrap of paper, underlined heavily, was the name of a hotel and a room number. Above the information I had penciled, "Man from Yucca Valley conducting experiments."

This was one of the men they said was conducting work far ahead of conventional scientists. I smiled knowingly to myself as I saw the predicted date for our meeting and noted it was near. I wondered if he had any pre-knowledge of our proposed meeting (later he confirmed that he knew he was to meet someone in the east but hadn't been told the name).

I also wondered what the fellow's name was, but soon learned. The next day after I found the note a friend telephoned excitedly.

"Howard, a man is coming to New York whom you'll want to meet. I can't wait to go to his lecture!"

I asked who the man was.

"George Van somebody. . . ." Then he got out a letter from a friend. "George Van Tassel. He's had contacts with the space people, and is head of the College of Universal Wisdom."

"Does he make experiments of some kind?" I asked.

"That I don't know . . . she (evidently the person writing my friend) doesn't say."

I figured this must be my man, and when the date of his arrival came I couldn't wait to get into New York and meet him. Bill Thompson drove me into the city, and we went to his hotel.

I called his room and checked my notes against the name. In a few minutes George Van Tassel stepped out of the elevator into the lobby. As he came into view I knew this was the man to whom I was directed. More exuded from him than merely a warm, pleasant personality. I could tell this man possessed deep understanding and a great deal of knowledge. In fact he reminded me of some of the space people I had met.

I introduced him to my party, but felt it unwise to mention what the space people had told me, while in the presence of the others, and looked forward to an opportunity to talk with him alone.

That proved to be difficult. First we attended his lecture, which was jammed to capacity. Afterward I tried to speak with him privately, but that was impossible because of the throngs which surrounded him. Before long I could see our conversation was generally public, and my pictures were being passed around for all to see. I discovered, however, that George and I were comparing notes telepathically, and I was firmly convinced he was the man I was seeking.

Within the week he came to my home, where he gave lectures to selected groups.

We appeared on the Long John radio show and the Steve Allen TV show—and the story really broke.

No longer did I have a private life of my own. Hundreds of people thronged our home in High Bridge, and from there on we had no peace in our household. Our visitors ranged from sincere individuals, to reporters, curiosity seekers, investigators—but I talked with all of them.

Suddenly during one of George's private lectures at my home I recognized someone who was to bring even more changes into my life.

As he spoke, I noticed a slim, attractive, young blonde woman raptly listening to the lecture, and I knew at once who she was.

My mind flashed back to a prediction the woman from Venus had made. She said I would meet a young woman who closely resembled the girl I had met years before, as a child, in the forest. She said her name would be "Marla."

I breathed the name . . . "Marla" . . . and the memory blocks began to disappear. She gave no sign of recognizing or noticing me. But I knew this was the girl who would work with me in the future, bringing once again into fulfillment a past affiliation on another planet—and a promise.

Then another realization overtook me, and I unwillingly pulled my eyes from her. This girl was not from another planet in this reincarnation. She was of the earth.

Apparently I was not the only one who sensed something different about the lovely young woman with the far-away look in her eyes. A friend from New York called me aside and asked if I had noticed her.

"There is something about her," he said. Then he repeated it: "Something about her, but I just don't know what it is."

Almost automatically I replied, "Yes, there *is* something unusual about her, and we will see a lot of her in the future."

At the same time I spoke, I knew that the work that I was to do on this planet would soon require a difficult personal decision.

It is a pity that many of us who are working here suffer from memory blocks which are unveiled only after we have already become established in a way of life—in my case, married, with children and the attendant obligations. It would take a great amount of understanding for one or both of the partners to re-define their lives along original purposes.

It was the visitors who assisted me in breaking through the memory block with which I was born on this planet. All of us have past lives. Some have expressed themselves on this planet in other bodies, other locations, but none of us originated on Earth. Thus life is an endless variety of continual growth.

In the beginning I had found a great many things the space people told me difficult to believe; but the more awareness I received, the more I was able to understand the truth of my own soul and its various expressions.

Now I was beginning to see into past lives and arrive at many answers to the questions I had often asked myself.

All of us have lived through hundreds of incarnations on various worlds. Some of us have volunteered to come to this planet and be reborn in earth bodies. We have volunteered to help in the work of helping other people receive the gifts of awareness. One can believe in these gifts on a purely intellectual level, defining the logic through rationalization, but believing is not knowing. To know a thing one must experience it.

Have you ever experienced some illuminating truth which flashed across your mind and gave an answer to a particular problem which had been worrying you? Or, perhaps, suddenly given insight into more profound subjects, and then, suddenly, the next instant, it was gone? Yet the memory of this brief illumination remained with you in your mind, but you can never find words with which to explain it to others. This is what I mean by *knowing*—it is difficult to put the knowing into words to those who do not know, or to those who only believe or imagine a thing to be true, without having experienced it. This gradual awareness and knowing come with evolvement of the soul, mind and body, which are ever striving upward toward perfection and oneness with our Infinite Father.

Once the memory block is broken, we do not need the constant surveillance or help from our space brothers. We are finally on our own. Only then we do experience a thrill even more satisfying than meeting and talking with these advanced people: the joy we receive as we help others achieve what we have achieved.

A few things I say will not conform to many of our orthodox religions. The occult sciences, however, do touch upon some of these subjects, yet many of these metaphysical teachings sometimes go off on tangents which rob the reality of every-day life applications from them. It may often be the aura of mystery many exponents of such teachings draw unnecessarily into them which are responsible for their failure to achieve practicality.

Science asks for proof, but how can we prove something which is beyond our sciences? Scientific proof is based on what we perceive with our five senses, not what we *know* with the use of even more valuable senses. But, somehow, out of these advanced ideas must come a science, a system of some kind; else our unchartered paths will become even wilder. Such a science, a cosmic science, which will involve investigations into the realm of other senses and dimensions, has already had its humble beginnings; but its patriarchs, like those ancient iconoclasts who announced the world was round and was not the center of the universe, have been persecuted.

Man's research has penetrated millions of light years into the universe, and into the almost inestimable depths of the atom; but now he must do research into an even more mysterious and difficult subject, MAN HIMSELF!

In talking of past lives, I should clarify the term, karma, as it has been explained to me. If you wish it, there is such a thing as karma, which one might term "A COMPENSATIVE HEREDITARY EXPRESSION IN THIS PHYSICAL THREE-DIMENSIONAL WORLD OF NOW." You can live above karmic conditions when the gift is yours, but woe to those who are slaves of karma, and cannot leap its prison walls!

To those who say they are resigned to live under this system of karmic retribution, let me say this: it is a false and artificial condition, one which takes on reality only because you agree that it should be so.

Jesus said, "Know the truth and the truth shall make you

free." Those who die on this planet without knowing the truth or having the gifts of awareness are prisoners of this planet and do not leave it; they are reborn in new bodies and continue in the school of life, bound to the wheel of karma.

Those, however, who die knowing the truth are not afraid to die. They know there is no death in Truth, and that they may be reborn on another planet, Venus, for example: a veritable heaven compared to Earth. Heaven is not an isolated place; instead there are degrees of heavens. Some souls who choose to be reborn here from another planet do so for a purpose, either to accomplish a mission or to be with some loved one. But whether they progress upward or downward, there is continual evolvement.

At the lecture I caught myself wondering if it was the soul of myself or Marla which had voluntarily cast itself once more into the hell of Earth. Perhaps it had been that of both of us.

CHAPTER 22

NATURAL COUPLES

Marla had so greatly resembled the girl on the rock I felt certain she must be the person to whom the Venusian woman had referred. I remembered our parting in June, 1946, when I asked her if we should ever meet again. She had answered in the negative; but added a statement which greatly puzzled me.

"No, Howard, but one will come who is my sister. She will work with you and be with you for the duration of this life span. She is my sister from Venus, and incarnated on this planet some years ago in your state of New Jersey. She is not too far from you at this moment. One day you will meet her."

"How shall I know her?"

"Don't worry; you will recognize her the moment your eyes fall upon her. Once in her presence you will know that she is the one of whom I speak. And you will discover she looks very much like me!"

At that time I had not envisioned any particular person; though I believe I pictured an Earth woman of mature years who might assist in my work and nothing more. A younger woman would conflict with my marital state.

Ten years later, while listening to George Van Tassel speak, I had seen Marla for the first time; and as the girl on the rock had stated, I recognized her instantly.

I looked at a silent, serious young woman with a sad face. My first recognition came not from physical appearance, but a feeling of oneness with the soul of the girl. And as I looked at her face carefully, my heart almost jumped out of my body: Marla bore a striking resemblance to the girl on the rock! She was shorter, and her hair, tied in a tight, severe bun at the nape of her neck, was not quite as light a blonde as that of the Venusian woman; her eyes were blue-green with flecks of gold, and the Venusian's had been golden. Yet my breath almost stopped when I noticed the similarity.

I enquired about the young lady, learning she had recently become widowed. Asking Van to come with me in order to give me courage to meet her, we walked over to her after the lecture, and I renewed an association with someone who had been very close to me in the past.

I remember her blushing when Van told her that she bore a scar on her upper leg, one of the ways in which certain people are "marked" and are known to their own as "one of themselves." Marla smiled and said, "But I have never seen a flying saucer, or had a contact with a space person." To which Van replied somewhat enigmatically, "You don't have to!"

Since I had first looked at her, my memory veil had been lifting. I knew I had known her before and had loved her and that we were meant to be one. While this was a happy revelation, it was also tragic, for I was already married. I was in a state of confusion, divided between fond anticipation and sorrowful foreboding of what would occur in my family life.

But after the first meeting Marla and I were irresistibly drawn to each other; and though both of us tried to fight off the predicted outcome, we were caught up in the overwhelming remembrance of a long-ago promise to each other.

I do not remember all my life as a Saturnian, but I recall being part of a family with parents, brothers and sisters. I was a spiritual teacher who instructed the young. I had at my disposal a space craft which I used for traveling to different planets for the purpose of both teaching and gaining knowledge. I taught many subjects, including the positive use of telepathic projection, and the study of God's Universal Laws. As such a teacher, I was known as one of the "Sons of Naro," "Sol do Naro," a teacher of Light who came from a region close to the Sun which was called Naro.

I should interject that the soul embodied on this planet operates differently, according to the frequency of the planet which makes up the magnetic lines of force around each corporeal planetary body (and all others for that matter). For instance, on Venus and Saturn the rate of vibration is much higher, and renders corporeal structures more tenous; and if an Earth man in physical body could go there he probably would not see some of the life forms which vibrate more rapidly than his own—no more than he can see the spiritual life forms in and around his own planet. Unless his physical body were processed and conditioned, he could not see the beings on another planet. If two planetary bodies are close in frequency, then, of course, the life forms are visible to each other. The life forms on Venus and Saturn, for example, are visible to each other, and their cultures are interchangeable because of the compatible frequencies. When souls incarnate on Earth, the frequency must be "stepped down," so to speak, in order to be embodied (born). Most of the time these reborns do not recognize each other, let alone know their own past histories.

On one of my trips as Sol da Naro I stopped off on Venus, and it was there I met Marla for the first time. Tall, lithe, with long, blonde wavy hair cascading around her shoulders, Marla, with her gold-green eyes, presented a picture far more beautiful than a storybook princess. We fell in love at once. As a Saturnian I was very tall, much broader than I am now. Yet there is a similarity in appearance; that is why this Earth body was chosen. And not only is there a similarity, but sometimes I actually do become the Saturian in height, size and powers.

Our love on Venus was intense and overpowering; but it was fated we should not stay together, since I knew I must travel to Earth and complete a mission which had been outlined from my day of birth on that planet.

I remember clearly now the day I left her. Both of us pretended to be very brave about it. Marla made little jokes and tried to laugh musically; but she found it hard to choke back the tears which crept into her laughter.

As I turned to look upon her for the last time, I made a promise to her. Someday, somewhere, I would find her again.

When I arrived at the portals of Earth a one-year-old boy by the name of Howard Menger had just died. The dead body was rushed to a Lutheran church to be baptized and prayed over. I, Sol do Naro, watched, and communicated with the soul leaving the little body. By consent and free will, and mutual agreement, I then entered the body. While the relatives prayed, the little boy miraculously "came back to life."

It seems strange, but I can remember the consciousness of the original soul, parts of its past (which was already impressed on the subconscious of the infant as soul record) as well as the partial recall of the Sol do Naro soul. As Sol do Naro I can vaguely remember being inside the craft as it hovered in the earth's atmosphere, then losing the sense of my surroundings and becoming as light. As this blob of light I entered the Earth body.

As my memory of the soul, Sol do Naro, became more pronounced through the contacts with the extraterrestrials, I began to operate less on the original subconscious record of the original soul and became more and more the Saturnian.

Although my original intent was to touch only briefly on

the subject of natural couples in order to save the embarrassment and pain of discussing the tragedy in my marital life, I have decided it best to explain the reasons for an unfortunate situation.

My first knowledge that something was happening both to myself and to my first marriage came shortly after I returned from the army, when Rose was heard to remark, "Howard is not the same man that I married." At that time I was not aware of my emergence as Sol do Naro, and subsequent misunderstandings pained me deeply.

When I had been quite young I had married the lovely, dark-haired girl whom I had met while working at Picatinny Arsenal. Our backgrounds were entirely different, which at that time made little difference to us—usually the case when very young couples first meet and share the novelty and spell of romance.

Parents, older and often wiser, know that in time, when the flush of romance dims, couples are faced with the down-to-earth and everyday prospect of living with each other. And at that time the basic characters and temperaments, due to background, training and other factors, finally emerge.

When we are young we cannot see these differences which stem from variation in mental and spiritual development. Our interest is usually based mainly on physical attraction and common interests of the moment. This is a story which is lived so frequently I need not go into it at great length.

On other planets, of which I have some knowledge, couples come together by natural selection; that is they know their proper mates. The considerations by which they choose mates involve mainly the state of evolvement of each individual. They choose their mates by a spiritual awareness of what constitutes perfect and complete union. They know a complete union must include all levels of development—spiritual, mental, emotional and physical.

In spirit like attracts like. As to the mental level, if the training has not been similar, then the capacities for mental development should be similar.

In the emotional and physical levels we first encounter the law of polarity, whereby opposites attract each other. That is why, as on Earth, an extremely virile man is attracted to a very feminine woman. We find this situation also operating in reverse. How many times have we observed a strong-willed, dominant, almost masculine woman married to a quiet, retiring little man, whom she "leads around by the nose"! We laugh at the situation, but we should admire! The gentle, often effeminate-looking man is strongly attracted to the woman who is physically and emotionally just his opposite—and the couple is happy. When two individuals are of like polarity, they repel each other, as do the same poles of two magnets.

If both members of a marriage union are of a strong, positive nature, there is friction and antagonism. If both are of a negative, receptive nature, always waiting for someone else to take the lead, then neither does anything, and the union lacks creativeness and progress.

The entire concept of proper mating is to bring out the best possible expression in both individuals. Together they should blend spiritually and mentally, but emotionally and physically complement each other and make a perfect unit of expression. Union and mating is not only a biological and social mechanism for the procreation and rearing of children, but also for the soul's development and fullest expression.

If a marriage does not fulfill natural mating requirements, the two persons are antagonistic, resentful, lack in creative interest in life, and contribute little to their own union, themselves, or their fellow man. They are better off separated than continuing an "armed truce" on the battlefield of an incompatible union.

This simple law—that in spirit like attracts like, and in the flesh opposites attract each other—is the basis for happy mating.

The space people know this law, apply it and teach it to their children, so that early in life their offspring know and recognize suitable mates well in advance of the formal selection.

Since evolvement is an individual process, differences in growth can occur, even on the other planets. When this does happen and there is a genuine difference in the spiritual growth and physical attunement, these people part with understanding and find more suitable mates.

Natural couples who find each other choose each other over and above everyone else. They stay together, not by law or by force, but by their own choice and are far happier than those couples who stay together because of law, social conditions or convenience. Thus natural selection is morally honest and spiritually valid. On this planet we rarely make natural selection of mates. But fortunate indeed are the souls who chance upon it. On Earth we have a word for it far more expressive, I believe, than do the space people. Here we call it "true love."

Children born of couples who do part on the other planets are loved and cared for by all. Children mature at a much faster rate; for example, on Venus a child two years old generally possesses physical and mental development enjoyed by an Earth child of seven.

Couples on Venus stay together much longer than Earth couples, if only due to the fact that they live much longer than we do. Their unions last hundreds of years and sometimes continue for several lifetimes. I believe this is as good a commentary upon their ways of selecting mates as one could offer.

In my own life, after several years of marriage, and particularly when my contacts became more frequent, differences in mental and spiritual makeup became more evident. Really it isn't anybody's fault when this happens to a marriage, and nobody should be blamed. There should be no cause for recrimination or bitterness; instead an attempt should be made to approach the problem in the light of understanding and to the best of all concerned.

I realize that in our society this is a delicate and extremely sensitive problem, and to date we have not adequately solved it. Perhaps if we teach our children how to select mates: first by spiritual attunement, second by mental pursuits and like interests, and third, by emotional complements, we need not teach them the fourth, for physical attraction of opposites is always in operation on the material level.

Sometimes it does occur that when individuals are of different states of development, the one who is more advanced will choose to remain with the one who is slowly advancing. Perhaps it is a case of previous commitment or karmic obligation; though often it is needless waste.

I would like to think that my wife and I parted as friends and with understanding. As humans touch each others' lives with purposes, sometimes we need part with purpose.

CHAPTER 23

CROSS-COUNTRY LECTURE TOUR

There are probably many reasons why a million listeners have taken Long John Nebel to their hearts. Some listeners will credit it to his fairness in handling controversial subjects and people; others his openmindedness; still others his personality.

The biggest reason for his popularity, however, is no doubt his capacity for bringing out the most interesting facets of the personalities he interviews.

I shall always remember my first appearance on the show, when I went to the studio with George Van Tassel. I was rather scared, not having been on radio before, but Long John soon put me at my ease, and Van's staunch support and sincere help gave me the courage I needed.

The biggest point of discussion was how I was able to take pictures of the spacecraft without any available light source. Van presented such an explicit description of how it was done that I here present his actual words, taken from a tape-recording of the show broadcast Oct. 30, 1956.

VAN: Well, John, this is further proof of what I have been trying to bring to the people for a long time. It is the greatest thing in our recorded history, and certainly in a free country it is the property of all the people. What Howard has brought forth here tonight (He referred to the photographs —H.M.) I consider, with my knowledge of the Polaroid camera and experience with the craft, authentic. I also consider them authentic for this reason: These were taken at different distances, and the perspective is perfect for the distances. The ship is emitting the same type of light as the one I contacted. The ship is luminous of itself. You don't need flash bulbs to take a picture of a ship at night. Now you asked Howard if this particular ship spun or seemed to spin and he said that he noticed a spinning action, or something appeared to be spinning. John, I would like to read from one of the messages I received telepathically to verify this particular thing he could not quite describe. This is in the book I printed on telepathic information, and dated August 24, 1952. "In the love and peace of eternal light, greetings to the mortal beings of Shan (Earth). I am Ashtar. Let me first inform you that we are grateful for your continual efforts in maintaining this contact power, for the information of your scientific minds throughout the planet, Shan, our ventlas do not spin. The emanation of spiral radiation from our ships give the illusion of spinning. The upper or positive polarity of the ventla radiates emanations outwardly from the center due to the collection and concentration of light particles through a vortice funnel in the center unseen. These light emanations radiating outward appear such as grooves on one of your phonograph records. The lower negative polarity operates in reverse. This light substance emanation is contained within a field of zero circumference, which is void, giving the impression of an egg. Your spectroscopic camera will reveal us only as light in the spectrum plus elements in your atmosphere. Advance this information also to those who still doubt."

LJ: Well, the only way you can find out whether there is any truth to any subject is to get all sides of the story. Now I know that one who talks about these subjects is frowned upon. I, myself, can't buy this story, and I am frank to admit it. I definitely agree with George Van Tassel, that this is not a man who has come out here to fool anybody, much less, he is not interested in fooling George Van Tassel, and I would say the same thing for Arthur Aho.

VAN: I know several million people have sighted these ships. There is a feeling of ridicule, and the fact that they won't report these to the military has created an unhealthy condition in this country, and I am bringing this out on the side of our government. I think we have the finest form of government in the world. I am talking for the government, not for myself, in that respect.

LJ: Why should we keep it a secret—what have we got to lose?

VAN: John, I have maintained from the beginning that the reason for the secrecy is the fact that if this principle of power gets out, it is so simple any average good mechanic could make a motor that would run—and it doesn't require fuel to operate. That, in my estimation, is the prime reason that this fact has not been revealed. That, and the fact that various information relative to that particular power and the things that have been found out since atomic energy has been gone into, has made practically every scientific book in our school today obsolete. They are behind the times. And the kids going to high school today read advanced things in *Popular Science*. And those kids don't object to learning, but they do object to learning things that are has-been material. That is the situation as I have analyzed it and I discussed it with many parents and many teen-agers. And this whole condition is bringing about something that is unhealthy for our country, because as I have said before, the strength of our country depends upon the people having confidence in their government, in their form of government, and the people responsible in government office. I just don't like to see the truth hidden. This is something that concerns everybody.

LJ: You use the word, "truth." You feel very confident that you know what you're talking about?

VAN: Positively confident . . . yes.

LJ: Howard Menger, you have been listening to George Van Tassel. Did you ever hear of George Van Tassel prior to his lecture in New York (October, 1956)?

HOWARD: Yes, I did, and that's as far as I would like to go.

LJ: (surprised) That is as far as you'd like to go?

HOWARD: Yes, I have answered the question.

LJ: Even George Van Tassel seems amazed at that one; am I right, George?

VAN: Well, I'll tell you, John; in dealing with these space people you find there have been a number of things that they have told various Earth people. I could not figure out how he (Howard) knew I was here in New York or how he knew where I was staying in New York, so I haven't had even the chance to ask him these things myself.

LJ: Well, I must say there is certainly an air of mystery about the entire situation.*

After this first broadcast as guests of Long John Nebel we had many return engagements.

In March, 1957, I headed west for a brief lecture tour, to be followed two months later by a month's tour of the entire Nation. During both tours, I stopped at Van's place at Giant Rock Airport, Yucca Valley, Calif. I also spent a few inspiring hours with George Adamski, a great soul who has reached an awareness of the truth of our being and our purpose. I shall never forget that warm, friendly meeting and his words of wisdom.

* Chapter 14 clarifies this mystery.

In April, 1957, Mr. Lester* and I made a cross-country trip, during which I had several contacts. The most interesting contact occurred in Hollywood, Calif. where Mr. Lester and I were staying at a motel on Sunset Boulevard. At 2:30 one morning, while he lay sleeping soundly, I was awakened by a strong impulse. I got up, dressed quietly, and went out into the foggy pre-dawn. I walked a few blocks, trying to "beam in" the impulse guiding me. I entered an all-night restaurant to have a cup of coffee and ask directions. I knew a contact was near, but I wasn't quite sure where or when it would take place.

I looked around the restaurant. Six or seven people were in various stages of devouring slivers of pie and cups of coffee. But sitting on stools at the counter, with their backs toward me, were a man and woman. The man, in particular, attracted my attention, for I was almost certain that a mental emanation came from him. I hesitated for a moment, then mentally asked him if he was my contact. He turned around, looked directly at me and nodded. So I went over to the counter and greeted them.

The woman moved over one seat and indicated I should sit between them. Sitting between these two people I again had the usual exhilarating feeling of love and friendship. The man was a pleasant individual, with brown wavy hair, a medium build, ruddy complexion and smiling eyes. I remember that he rather reminded me of John Garfield. The woman also had wavy brown hair. Her eyes were dark, and her skin white and camellia-like. She wore no makeup except for the faintest suggestion of lipstick. She was small and slender, with a sweet, gentle face. She did not speak at all.

"Howard, I trust your companion is sleeping well?"

"He is sleeping like a log, and I am surprised he didn't wake up when I got up; because he is a light sleeper."

I was reminded how that whenever I had a mission to do I could always leave my house without attracting attention. I wondered why the rest of the household always slept so soundly.

We spoke about my trip and the lectures, and he asked me how I liked California. I told him I liked it very much and would someday like to live there, because some parts of it reminded me of Venus.

Then, as if he had overlooked it, he smiled apologetically

* "Mr. Lester" is a pseudonym.

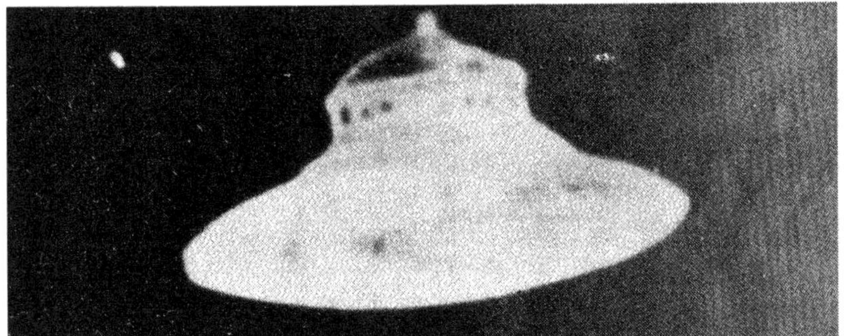

Venusian reconnaissance craft hovers over field location No. 1. Polaroid photo taken by the author.

Venusian man allows the author to photograph him in silhouette against background of lighted spacecraft. Space people are reluctant to permit clear photographs of their features, because they might be recognized while mingling with Earth people.

Venusian ship hovering two feet above the ground.

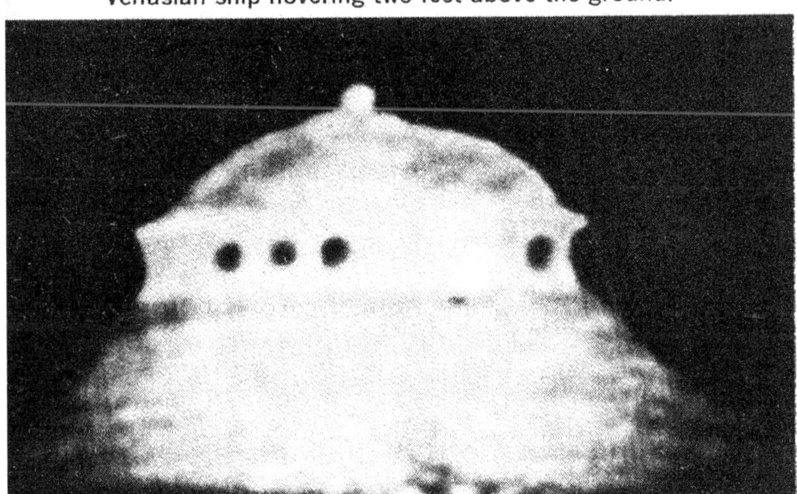

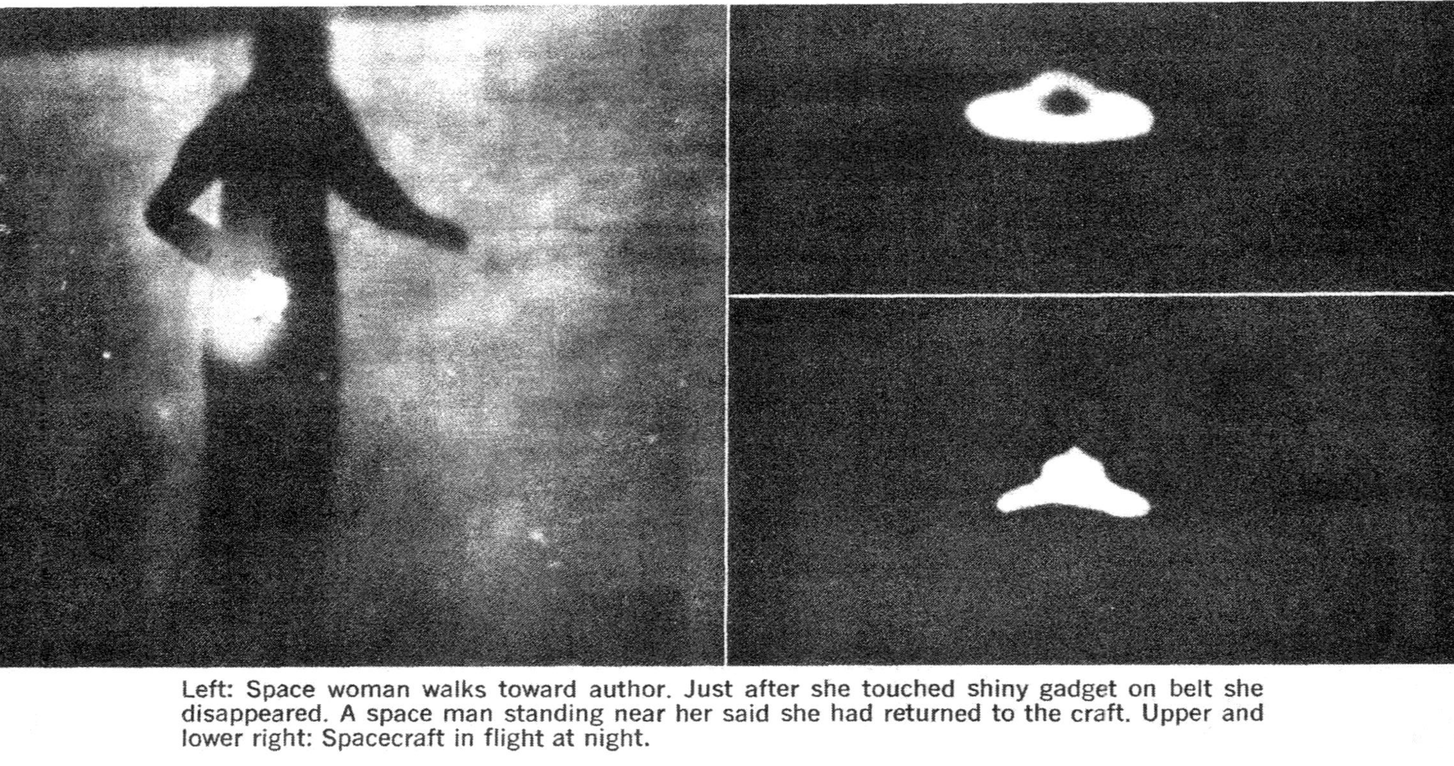

Left: Space woman walks toward author. Just after she touched shiny gadget on belt she disappeared. A space man standing near her said she had returned to the craft. Upper and lower right: Spacecraft in flight at night.

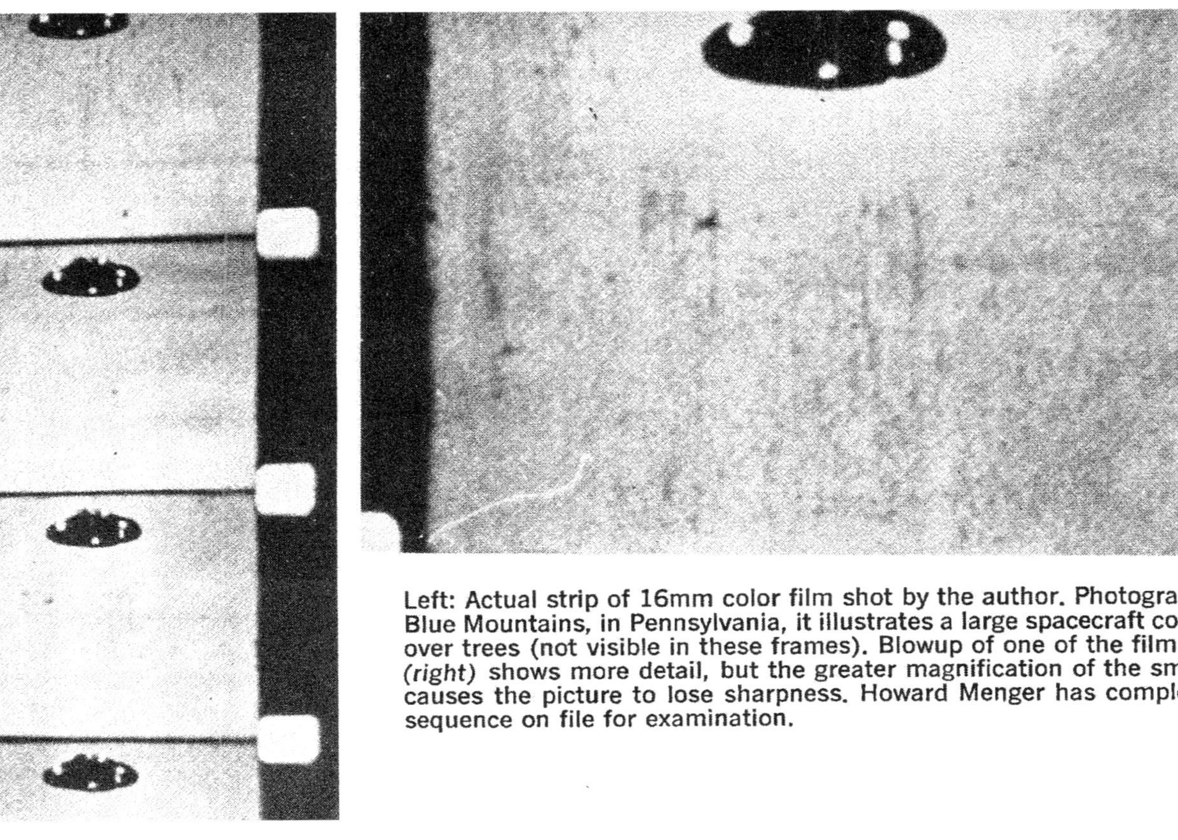

Left: Actual strip of 16mm color film shot by the author. Photographed at Blue Mountains, in Pennsylvania, it illustrates a large spacecraft coming in over trees (not visible in these frames). Blowup of one of the film frames (*right*) shows more detail, but the greater magnification of the small film causes the picture to lose sharpness. Howard Menger has complete film sequence on file for examination.

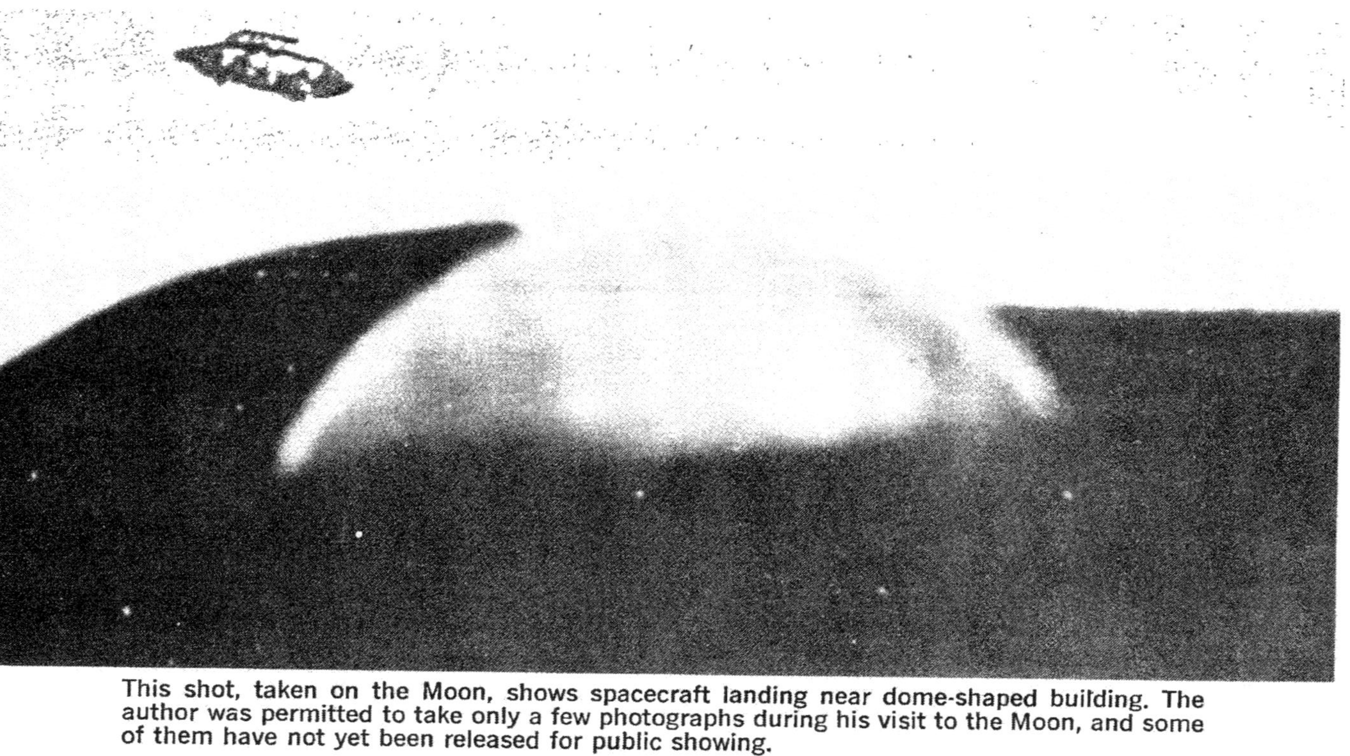

This shot, taken on the Moon, shows spacecraft landing near dome-shaped building. The author was permitted to take only a few photographs during his visit to the Moon, and some of them have not yet been released for public showing.

Two scenes from motion picture filmed by author at Field Location No. 1. These two pictures, shot from a movie screen in close sequence, show a space ship coming down and landing.

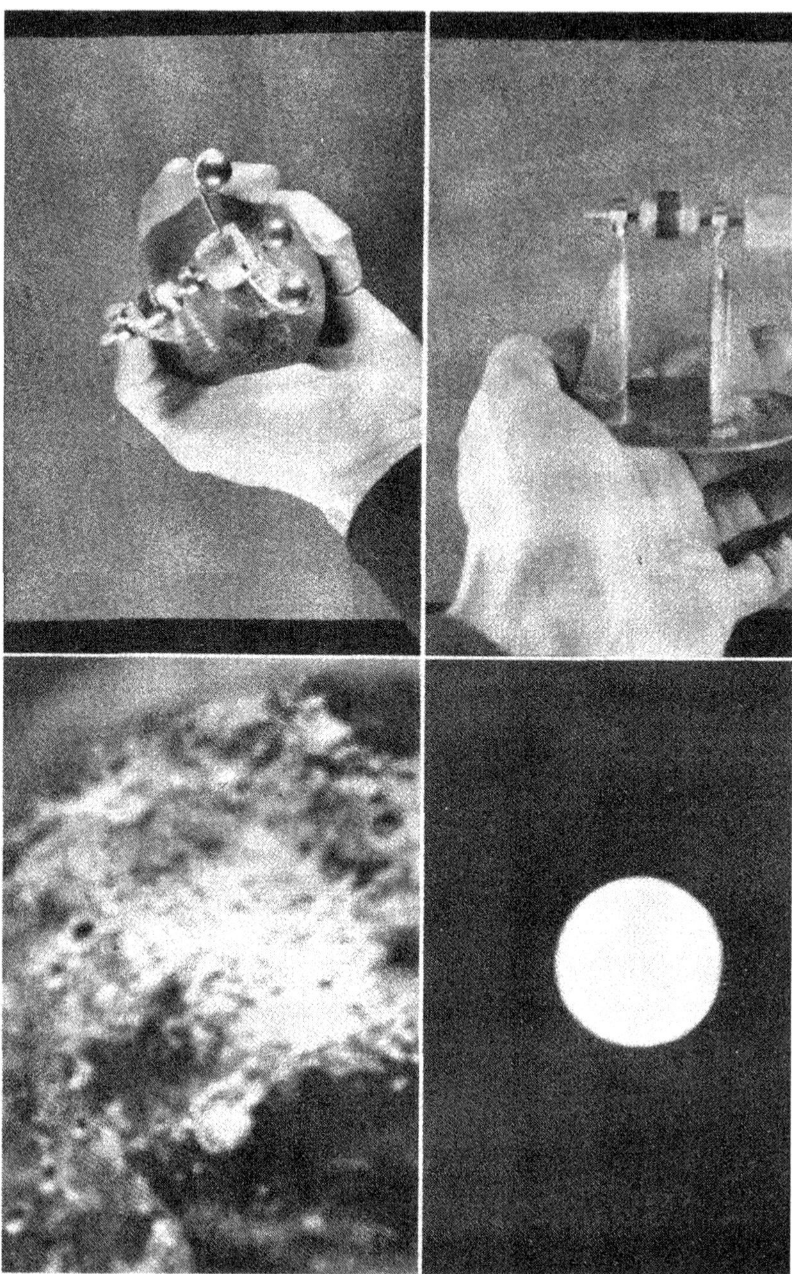

Top left and right: Two views of a free-energy motor the author constructed under the direction of the space people. Bottom left: Closeup of Moon's surface showing atmospheric clouds. Bottom right: Author took this photo of the Moon as he approached it in spacecraft.

Craft hovers in foreground while in the background is pictured one of the Moon's mountain ranges. Bottom: Closeup of typical lunar terrain. The odd protrusion at the left is an artificial construction.

Howard Menger.

Printed in U.S.A.

and said he should introduce themselves. He was Suna and his gentle companion, Karma. Both were from Mars.

"I am known as Suna by the contacts on your planet, but as you know we are not known by such names on our own planet."

I was surprised and somewhat suspicious at hearing the names, for this was the first time I had been given actual names. Maybe these people were only posing as contacts! Anybody could act like a space visitor, especially if he had the ability to communicate telepathically.

He immediately picked up my doubts; smiled; and then my eyes were drawn toward his. As I watched him his right eye turned to a very light gray, then back to brown! Then he did the same thing with his left eye.

I must say that thoroughly convinced me; then I reasoned that by having names, space visitors could be easily identified and referred to by other contacts. It would facilitate our work.

Sensing I accepted him, he explained he had been on this planet for some time and had taken on the complexion of terrestrials. But I saw a look of pleasant expectancy on his face as he said he soon would be returning to his home planet.

"I have followed you all across the country," he told me, "and will continue to do so. When it is advantageous I will make myself known to you and to witnesses."

He said a photograph would be taken in which he would appear with a crowd of people, and that I could use the photograph if I wished for my book.

We discussed my group back home, and he suggested I stop "pushing" them too hard, for it was difficult for them to understand some of the concepts. I had begun teaching by briefing them on the space people as to where they came from and their purpose. Then they had a brief course in Scientology, recommended by a space contact. Next, I gradually introduced them to the study of metaphysics. I hoped to make teachers of some of the group so they would be able to have their own contacts and carry on the work.

Suna suggested it might prove well worth the time for our group to study "Self-Realization," as taught by a great teacher, Parahamsa Yogananda. That was where Mr. Lester, with his knowledge and devotion to the study of Yoga, could greatly help us. I should allow him to proceed with the teachings in his own way.

Abruptly Suna said, "And while you are up there on the platform at Giant Rock, you can take my picture."

This came as a shock to me, for I had not anticipated attending the spacecraft convention; however, I had learned always to follow their suggestions. So Less and I headed toward Yucca Valley.

While I was on the platform giving a brief talk, surely enough I spotted Suna in the audience, asked for my Polariod camera to be reached up to me, and snapped a picture of him in the audience. I hoped he would be visible in the finished print.

On the way back to New Jersey, Suna and Karma made their presence known in various ways, and I was happy to know they were always near. One time Suna left a sign on the entrance door at a lecture hall in a Chicago hotel, and I called it to the attention of John Otto, who was sponsoring my appearance.

In the summer of 1957 my son Robert left this earth. He was brave, cheerful and brilliant to the last. A few months later my Dad passed on. Within two years there had been four deaths in my immediate family.

Everything in my personal life seemed to be coming to a head. Business dropped off. The manager of a large beverage company, one of my largest accounts, called me in to tell me that because of my story of the space people my services were no longer wanted. If at any time I felt ready to quit, that was the time.

I had to make a decision: whether to conform and appease, or to devote all my time and energy to spreading of the knowledge. I chose the latter. When I gave up the sign-painting business, my financial position became precarious. Yet, in the back of my mind, I knew that with positive thinking I would be able to solve such earthly problems.

To be almost alone in a crusade was not easy. My concepts and understanding had been expanded with the wonderful knowledge I had learned from my space friends. But to interpret those concepts in our society was difficult. Many unpleasant situations arose, due to the misunderstanding and misinterpretation of my acts.

When I started to spread the message of love I meant a sacred, spiritual love, for a person or persons. A spiritual love which expresses through that one small spark of the Infinite Father present within each of us. This is an impersonal love expressed to all.

Some experiments I attempted with the groups and with certain individuals in the group failed; yet some succeeded.

At the end of 1957 I gave the group over into capable hands and secluded myself so that I could study, put down my experiences in this manuscript, bring forth the music that had been given me, to go through further training for the even greater work ahead.

CHAPTER 24

"PROJECT MOON"

As my space friends had promised, they took me on my first trip to the moon the second week of August, 1956.

It began with my receiving a telephone call from a man in Pennsylvania. I had been working with this man for some time. He was a well-known citizen in his community, a successful business man who donated all his spare time to establishing contacts, disseminating information, and assisting the space people in their work between the earth and moon.

When I picked up the phone and heard his voice I was impressed, for I realized he was a very important person among those carrying out the work on Earth. I remembered how one time he had asked me for a list of people who, in my opinion, would be good contacts, dependable and trustworthy. I eventually gave him such a list, then learned two of the people on the list had been contacted by the space people.

One of the two people had talked to me considerably about his contact. This individual had been given samples of processed food from the moon which he had kept in his possessions for eight months, not being sure what he should do with it, and asked my advice. I suggested that we arrange to have one specimen, a processed potato from the moon, analyzed by a reputable laboratory. So we went to the largest laboratory in Philadelphia and left the sample with them, then waited for the report.

We classified the sample as an "unknown," though later we told them what it was. Their analysis came back as follows:

Total weight of sample	5.20 grams
Moisture	7.23%
Ash	4.49%
Fat (ether extract)	0.95%
"N" as Protein (NX 6.25)	15.12%

Reading the report, we were immediately struck by the most unusual point in the analysis: our own potatoes grown on Earth do not contain more than two or three per cent protein!

Next, we decided to obtain a carbon-14 test to determine the age of the potatoes, since we had been informed these processed food products could last indefinitely. My wife, Marla,* and I were going to the west coast to do some research for a few months, so I left the specimens in charge of the individual who had obtained them. When we returned we learned no report had come from the lab. We decided to go to Philadelphia and talk to the man in charge of the analysis.

He informed us they had not proceeded with the carbon-14 test because they needed a written authorization to do so, and that the cost of the test would approach $2,000. At that point we were glad they had not proceeded, for while we had paid for the previous test, we could not afford that large a price.

The doctor, very polite and interested, suggested I take their lab report and the specimens to a government agency (whose address he gave me), and volunteered that if any of the scientists of that agency had any questions about the previous analysis, he would be glad to cooperate with them. We thanked him and left.

When we returned home we immediately called up the individual from whom I had obtained the sample, and who had not been able to make the Philadelphia trip with us, asked him to go with us to the agency the laboratory had recommended. Two days later we went there, where we were advised as to which of their research labs we should take it. Finally we arrived at the lab itself, where we talked to Dr.——, a polite, intelligent man who appeared to be completely fascinated with the specimens presented him. He assured us he would immediately get to work on a thorough analysis, and that there would be no charge to us. We left the samples with him, feeling we were on the right road.

About two weeks later, my wife, my friend and I went to the lab to check on the progress of the analysis. We were shown into a room where a piece of a specimen was in a container of water, another piece was in another container, and a small fragment was under a huge microscope. We took turns looking through the microscope.

Through the microscope the outer surface of the specimen appeared like a crystalline beach of sand. It was beautiful to look at. My wife suggested the crystalline effect could be due to an intense contraction of structure. I then told him what

* Marla Baxter and I were married in May, 1958.

had been explained to me about the method of collapsing the molecular structure, which probably gave the potato a dehydrated nature.

My friend suggested they plant one of the specimens in the ground to see if it would grow. We were told they would run all sorts of tests on the specimens left with them and keep us informed, and that we were also welcome to come to the lab any time we wished to watch the progress of the experiments.

Since my friend had been given the potatoes in the first place, we left the arrangements in his hands. That was June, 1958, and the last we heard from the laboratory. I understand that it has now become classified information, and the reason I am including this is because I believe such knowledge belongs to all the people—not left in the hands of a few.

But back to the man responsible for my friend's having the potato in the first place, the businessman from Pennsylvania.

At his suggestion I met him at a small restaurant off Route 22 near West Portal, N.J. He gave me a date and time to go to a certain location in Pennsylvania, already familiar to me.

A few days later I went to Besecker's Diner (Route 611, in Pennsylvania) and again met my friend. This time he was accompanied by a young couple. The man appeared to be about 20 years old (later I found he was 79!), was of average height. His hair looked as if it had recently been cut to a shorter style. His woman companion wore her strawberry blonde hair shoulder length and had a faint trace of lipstick. She, I later found out, was 69 Earth years old.

They told me they had been on Earth one week, having come here from Mars; yet they already spoke English, Russian and German fluently. They explained they had learned to speak the three languages by the aid of a device which I can best describe as an ampli-converter. The young man explained it something like this:

The would-be student of a new language is seated comfortably in a chair facing a view screen. An electrode is connected to each temple and wires run from the head to a small instrument about the size of a portable radio. This instrument is an electric memory brain which contains the electronic thought patterns expressed by the words through sonic vibrations fed into the machine by a previously recorded voice speaking the language and at the same time thinking of what the converted words mean. By this ingenious device the student is able to learn any language quickly, because the visual and audio transmissions are equally impressed on the brain cells; in addition the electrodes connected to the temples impress the memory cells of the brain so that it retains the ability to speak and understand the language.

The young (?) man then told me I myself could conduct experiments which would at least demonstrate the process and prove that brain impulses, as related to our thoughts, can be picked up through the use of electronics—and with an ordinary six-tube radio and a tape recorder!

At a later date in the home of a friend, Art Aho, of California, our host, Mr. Lester and I tried the experiment and did prove to ourselves that brain impulses could be changed to electrical energy and then to sound energy. Mr. Lester agreed to be the subject, so we hooked two electrodes to each temple, then made connections to the loop antenna of the radio—one to the ground, another to the condenser. Then we connected the tape recorder into the somewhat odd circuit.

We looked at Mr. Lester sitting there with the electrodes attached and Art asked him very sanctimoniously, "Do you have any last words." Then I burst out laughing as I saw the ludicrous similarity to a formal electrocution. But Mr. Lester's unperturbed and serious manner got the two of us serious again and I instructed him to concentrate upon a constructive line of thought. We clicked the recorder to "record" position and waited. When we thought we had recorded a long-enough sequence we reversed the tape, then played it back.

I suppose we had half expected to hear a voice similar to Mr. Lester's on the playback; but instead we heard sounds very similar to teletype transmission code one can hear on the short-wave band of a radio. We assumed the sound came from the conversion of the brain impulses into sound.

I have purposefully left out certain details of the hookup so that this experiment will not be done by amateurs. For if the chassis of the radio were tapped in a "hot" place, the experiment could cause serious electrical shock. They had specified a Philco six-tube radio, incidentally, for the experiment; however, we tried it with a console type radio, and that may have been why the results were not as good as we had expected. But we did get results. Later another contact suggested I carry on further experiments along this line with an electronics engineer and to try and establish a method of changing the sound energy (as received on the tape as an audible teletype-like code) to light energy by way of a screen, thus making it possible for one to see his own thoughts!

But I have digressed again, and now let us get back to the meeting with my friend and the young space couple at the restaurant.

After finishing eating we went out, got into my car and talked further. They mentioned three men who were slated for work on "Project Moon," as they termed the work we would do. This would involve receiving samples of processed extraterrestrial food, plus foods from the earth processed on the moon, though at the time I did not understand the significance of the work and wondered where its value laid. But I was so overjoyed at the prospect of making a trip to the moon, I asked few questions at that time!

Two of the three men they mentioned would not work out, because of the attitude of their wives. If they found it necessary to be away evenings without sufficient explanation —and they wouldn't be permitted to explain fully—their families could conceivably be broken up.

The man paused in his talk and rolled up the windows of the car, as I wondered about his unusual action: for it was a very hot day.

"Just for a minute," he apologized; I noted his tone was becoming more confidential.

"Howard, as far as I know, you've been shielded from what I now have to tell you, but it is necessary for you to know. You're becoming too widely known and it may happen to you."

I tensed, worriedly. This was the first time I had been spoken to in such fashion. The man sounded as if some danger threatened.

"Any of these men working with us, including you, may be approached by false contacts. They may be able even to present you with authentic specimens of processed food."

"Who are these people? I thought I could listen to any of you!"

"They're not US, Howard. There are OTHERS operating. I'll speak of them simply as 'The Conspiracy'."

I was further shocked and taken aback as they described a certain man, possessed of so much knowledge of space activity and such an operator of deceit, he would confront us with extraterrestrial specimens and might even arrange a trip to the moon—while all of this would be backed by a libelous story against our space brothers.

"Many gullible people will be unwitting helpers to this conspiracy (I, too, would on one occasion become victim to the belief that I had contacted a space person and later find out this was not so)."

They described the operator to me. A man about 40, with brown hair, graying at the temples. Average height, about 160 lbs. He smoked cigarettes, but occasionally cigars, usually wore dark brown suits, black shoes, drove a late model car and lived in the vicinity of Somerville, N.J. While reading he usually wore dark-rimmed glasses. His movements were described as quick and alert.

I wondered as I listened incredulously. If this were true, how could we determine who was our friend and who our enemy? I could see friends divided by suspicions and reluctance to trust former confidences. With my mind in a whirl, I thought, "Suppose one were contacted by an alien and were sworn to secrecy on the promise of future contacts and trips in space craft? How could one know he was not being contacted by a real space brother?"

As a matter of fact, could I really be sure the people I was then talking with were representatives of the real space brotherhood who wished humankind well?

The man looked at me sadly. "My friend, this earth is the battlefield of Armageddon, and the battle is for men's minds and souls. Prayer, good thoughts and caution are your best insulation."

I shouldn't have doubted these people for a moment, but I was quite ill at ease. I had been sheltered from the knowledge that all of the space people's work on this planet is not sweetness and light. The others I had contacted must have been on the "right side"—for what I had seen of their mode of travel, their way of life, their courtesy and good will toward each other and all mankind, convinced me they were a good people.

Then the young lady spoke: "You don't know, Howard, that there is a very powerful group on this planet, which possesses tremendous knowledge of technology, psychology, and most unfortunate of all, advanced brain therapy. They are using certain key people in the governments of your world. This group is anti-God, and might be termed instruments of your mythical 'Satan'."

I listened, aghast. I remembered that at one of my lectures

a little old woman, wild and strange looking, had motioned me aside. She had whispered into my ear, seemingly afraid she would be overheard, something of "dark forces," which were "everywhere." I had remonstrated that the space people were good, and that was that, feeling the unusual little lady had lost her faculties. In the light of what I was hearing, the recollection of the old woman's tale became frightening. When I felt I knew everything I had indeed become, in some respect, an overbearing sophist, the type of attitude I had hated when I saw it in others.

"They use people not only from this planet," the young lady continued, "but people from Mars as well. And also . . ." and she looked at her companion; he nodded his assent.

"Also OTHER PEOPLE OF YOUR OWN PLANET—PEOPLE YOU DON'T KNOW ABOUT. People who live unobserved and undiscovered as yet. It is a kind of 'underground,' in your popular terminology. This group has been infiltrating religious organizations to dupe your peoples into a distorted concept of a truth which enveloped your planet thousands of years ago. They are using the credulity and simple faith of many people to attain their own ends."

For the first time there was anger and frustration in her voice.

CHAPTER 25

ORBIT

Apparently that was all they were going to tell me about the unpleasant subject, for the young couple got out of the car, went to their own car and indicated we should follow.

We drove out Route 115 toward Mt. Effort. When we approached the Blue Mountain area, we turned off onto a dirt road and I recognized a familiar spot. I remembered that a transmitting instrument had been placed nearby some time ago. My friend got out of my car to inspect the instrument and I followed. I knew the instrument was tuned in to at least 50 individuals, three of whom lived in Allentown, Pa. Their brains were acting as transmitting units for the instrument, although they were unaware of the fact and in no way affected in their health or free will.

That business effected, we got back into the car and followed the lead car as it again pulled out onto the road. We drove for about three miles, turned off on another dirt road and traveled for three more. We approached a clearing where I made out two craft with a small crowd of people standing around them. At the entrance to the field stood a young man waving an instrument which emitted purple light and which I suspected registered our brain impulses.

We left the cars and approached the man waving us in.

"Hello, you're late. Ten minutes later the tower would have been dismantled and all of us gone."

We apologized. As we walked toward the smaller craft we noticed 10 or 15 men who began to dismantle a 10-ft. structure which was surmounted by revolving red lights. By the time we reached the craft, the structure had been completely dismantled into sections and was being loaded aboard the larger ship. I marveled at the rapidity with which they had accomplished the work.

We mounted the ramp, entered the craft, sat down at a large round translucent table. I looked out and saw the ramp being retracted as the iris-like doorway closed. Then we heard a high-pitched humming sound as the craft prepared to take off.

I still wasn't sure where we were going. The young lady, reading my thoughts, informed me, "I'm not absolutely sure, but I feel certain we're going to orbit the moon once or twice."

The lights dimmed as the man at the controls pushed some buttons and the view screen lighted with a view of the moon's surface, which I guessed was about 25 miles away.

"Are we there already?" I asked, remembering our quick trip into space when we had observed Venus.

The young lady laughed. "Howard, we haven't even left the ground yet! We can tune in and observe any planet in closeup with our instruments."

I watched the screen as it momentarily went dark, then tuned in a closeup of what I presumed was a crater. As the fuzzy image was brought into sharp focus, I could see blues and greens showing in the darkness of the crater.

"How is that our astronomers can't see these colors?" I asked, and she answered: "Your instruments are limited because you use light which is of a secondary energy and reflected off the atmosphere, which, in turn, reflects an effect of atmospheric distortions. We do not use this type of energy. Our source of energy, though it be secondary, is without distortion. We can pick up and even photograph forms in your own atmosphere of which you have no concept.*

As the view screen then focused away from the crater and over the general terrain, I saw dome-shaped buildings coming into view and a craft making a landing near one of them. At this instant I felt the slight jerk sometimes associated with taking off and knew we were on our way into space.

The view screen darkened. A sliding panel opened in the wall and two people entered—Earth people, I noted with surprise—led by a space brother. I was further surprised at recognizing them, for they were people I knew. They joined us at the table.

There were now nine people in the craft—eight at the table and one man at the controls. We greeted the newcomers as the room darkened again, the view screen again lighted and we saw the moon once more. We seemed to be approaching it at a tremendous rate of speed. Now and then I saw the now familiar objects approaching us and

* Trevor James in his book, "They Live In the Sky" (New Age Publishing Co., 1542 Glendale Blvd., Los Angeles, Calif., $4.50), claims to have photographed animal-like forms in the sky by the use of infrared film and filters, and presents the pictures in the volume. Could this be what the space people referred to?—The Publisher)

veering away, but I knew the meteors could never strike our craft due to the shielding field.

The moon, of which we had seen a relatively small image, had grown larger and now filled the screen. Land masses began to appear and the terrain became more distinct. I asked how fast we were going, and the pilot told me 82,000 miles-per-hour.

"We will decrease or increase this speed according to the density of a shower of what I will term 'objects' ahead of us."

"Meteors?"

"No, not exactly. They represent matter in the form of different elements present in space which eventually form planets under the control of a funneling magnetic influence. These and meteors will present a hazard to future expeditionary rockets from your planet unless you discover the secret of repelling this material."

The others were not asking questions. Instead they were deeply engrossed in the picture of the moon's surface. From watching the screen I figured we had traveled around it at least twice, for I had seen certain terrain revolve past the field of view twice. At one time we caught sight of Earth in the distance, glowing bluish-white with tinges of red, floating like a tennis ball in an inky black pool. As I saw the earth I grabbed my camera and waited until I caught sight of our planet through a porthole and quickly got it in the viewfinder. I also tried some shots of the moon. Of the five photographs, only three came out, and they were blurred, but to me they are indeed priceless and proof that we had orbited the moon.

"You're on your way home," the man accompanying the young lady announced, and the view screen switched to a shot of the earth approaching rapidly. When we landed at the field from which we had embarked, I checked my watch and found we had been gone only two hours.

We parted company, each of our cars going separate ways.

"Don't look so disappointed, Howard," the young lady said as we left them; "perhaps the next time you'll land!"

CHAPTER 26

DESTINATION LUNA

In September, I again met the same agent at Beseckers Diner, the meeting again arranged by a telephone call. This time he was alone. After having coffee we left in my car and drove to the same field from which we had embarked previously.

"Well, Howard," my friend announced as we neared the location, "this time I think we're going to land on the moon. If we do, you will have a great experience."

One craft was waiting for us. As we entered it I was again surprised to see people whom I and members of my Thursday night group knew personally. One of them, an elderly man, was not, however, a member of any saucer group, but a fellow of great prestige in his community. I knew personally that at one or more times in his life he had been persecuted by orthodox agencies of conformity. I was so moved with emotion in meeting my old friend that I actually burst into tears. His eyes filled with warmth as we greeted each other; then after greetings were exchanged all around, the craft took off—Destination Moon.

This time only six people were aboard: one space visitor at the controls, one at the table acting as instructor, and four people from Earth.

The man at the controls began speaking with a slight accent through a loud speaker:

"My friends, this trip will be a longer one than the previous journey. You will soon go through a processing, whereby a complete change will take place in your atomic physical system. Each atom of your physical body will undergo a processing which will change its polarity, frequency and vibration, to adjust your body from its balance to the earth's attractive inertial mass to that of the moon's. This will require approximately a week and one half, Earth time. Do not be alarmed at the initial effects. Nothing can harm you. Now keep your eyes on the view screen."

We looked at the screen and saw the earth fast diminishing in size as we sped away from it. Again the voice reassured us:

"Please do not be alarmed. Remember that we are only expressions or projections of a reality, which in truth does not exist. You are being changed atomically to fit a reality of expression, or what you call the moon. When we reach it, my friends, you may observe and take pictures at will; but the processing will continue as we orbit the moon and will seem to involve a week and a half of your time."

The lights dimmed and were supplanted by a yellowish glow over the entire room; then the lights came on again, and I felt a strange sensation. For a few seconds breathing seemed difficult, but then it became easier and easier. It is difficult to describe my physical and mental feelings. I seemed to think more clearly, to compose my thoughts and reach conclusions more quickly. The senses seemed to be stimulated. Colors became more vivid; the sense of smell became sharper, for I remember becoming aware of perspiration odors of myself and companions. My sense of touch must have been accentuated, for I could feel the seat under me more distinctly, though it in no way made me uncomfortable. The man at the controls explained the processing was now being carried out and would be continued.

The instructor sitting with us saw one of us stifle a yawn. He laughed and mentioned something about it being our bedtime. For the first time I wondered just where we would sleep.

He led us through a doorway into the sleeping quarters, which were arranged in sections with three beds, bunk style, one atop the other, to each compartment. My elderly friend, myself and another man were assigned to one of the sections.

I said, "Well, we might as well try them out," as I climbed to the top bunk and stretched out.

My elderly friend sat on the bottom bunk and felt the softness of what must have been a mattress. "Ahhh!" he exclaimed; "this is really something!"

I removed my outside clothing and put them on a kind of built-in hanger, set my shoes on a ledge. The bed did not feel overly soft; instead it seemed to give just the correct extent to support the contour of the body. I laid my head on a flat pad of soft material, pulled the single warm (but extremely light) coverlet over myself. Despite the excitement of the day I fell asleep quickly.

We were awakened by a gentle knocking on the sliding compartment door. It was our instructor who said it was time to get up. I looked at my watch, and discovered we had slept only four hours; yet I felt rested and refreshed, as if it had been eight or nine hours.

My first reaction upon arising was to go to the porthole and look out to see where we were. Various blobs of light of different colors filled the field of vision, along with one giant red ball, which looked like a huge planet. Later I was told it was the sun, though I never understood why it did not appear bright.

Next on the agenda was a warm, invigorating shower in a compartment containing three or four cubicles, partitioned by translucent walls. When I stepped inside one of the cubicles the door closed behind me and lights went on automatically. I saw automatic controls for room and water temperature. Three shower heads, one above me and two at waist level, could be operated separately or all together.

I pushed a button and a flow of water, apparently mixed with warm air for it was quite bubbly, fell over my body. I had never had a shower so invigorating. I looked around for soap, but there was none. Seeing another button I had not previously pushed, I put my finger on it and a stream of a colorless solution came from the shower heads and completely lathered my body; at the same time the water was turned off. I pushed the "soap" button, then the "water" button, alternately, enjoying the novelty like a small boy would have done.

I could hear my friend trying to sing in the next cubicle and assumed he had mastered the technology of the shower, though fortunately, for my musical sensibilities, the partitions were partly sound-proof.

"Howard," I heard him call faintly, "this thing had better be what I think it is, for I'm going to use it!"

The sanitary facility to which he obviously referred looked very much like one on Earth, except that the bowl was lower to the floor and was made of a hard white translucent material, not a ceramic. A kind of sink with the same bubbly water was a part of a built-in vanity arrangement, complete with mirror. I looked at my face, thinking I would have to borrow a razor from someone. Surprisingly my beard had not grown—and throughout the trip we found it unnecessary to shave.

I stepped out of the cubicle into the main room and joined two others already waiting there. Then I smelled food and suddenly discovered I was ravenously hungry. Our instructor opened a compartment in the wall and withdrew some items of processed food, which he put into a deep well, or pot, set into a sink-like unit. He pushed a button and the pot filled with liquid. He allowed the food to seep in the liquid for about five minutes, then he drained the liquid from the pot. He pushed another button and almost instantly the appearance of the food changed and steam rose from it. It had been cooked in little more than a second!

"You will forgive me," he said apologetically, "if I do not don the traditional caps sometimes worn on Earth during such operations, particularly the backyard amateur kind of barbecue cook."

He removed the food from the pot by means of a large, deep strainer and transferred it to plastic-like plates which he said were disposable. He set the plates on the table. "No, I haven't forgotten your juice," he laughed, and drew fresh juice from a spigot in the food compartment wall.

During the trip we enjoyed many kinds of processed foods, including cabbage, parsley, carrots, potatoes, very large wheat kernels, kernels of corn. We used a green mineral salt to season our foods, and I particularly enjoyed some kind of spread, similar to avocado butter, but white in color.

Often we were served nuts from other planets, though only the nutmeats and I didn't have a chance to see the shells. One of them, large and almost a meal in itself, was served in slices. Another kind tasted something like a brazil nut. I remember eating with great pleasure a fruit which was about six inches in diameter, round, smooth, orange to red in color, with skin like a nectarine. When one bit into it he found it very juicy. It tasted like a combination of peach and plum. The pit was small, and looked something like a plum pit.

All of the vegetables were very tasty. The potatoes had a meaty, nut-like flavor, probably because of the high protein content. The parsley leaves were much larger than our variety, but less strong in flavor.

My friends who read this will probably laugh at my talking so much about the food, to the exclusion of many other interesting things, for they know what an enthusiastic eater I am. However we passed our conditioning period away in many other pleasant ways. We listened to music which came from the earth and other planets as well; and constantly we talked to our space friends, learning much from them.

The view screen proved to be a constant source of interest and delight as we tuned in on different planets and saw scenes of these fascinating worlds. We communicated through the view screen with other craft, and other agents in different locations on the earth, the moon and elsewhere. There wasn't a dull moment.

I am not certain how long we spent in these pleasant occupations, but, estimating by my watch, I believed it to be about ten days. I have often thought that time might have been different, possibly because my beard didn't seem to grow; but that could have been a result of our conditioning—however, all our other bodily functions seemed to progress normally.

Although there weren't any "No Smoking" signs and I assume it would have been all right with the crew, I noticed nobody smoked during the trip, or later on the moon. For the first time in years I had no desire to puff on my faithful old pipe. Again I reflected that time seemed to stand still, and yet there was constant activity on board the craft. I am still perplexed, and confess that the matter of time is still too complicated a subject for my becoming intellectually involved.

CHAPTER 27

SIGHTSEEING TOUR

Finally came the long-awaited announcement. Through the loud-speaker the man at the control panel informed us we were preparing to land on the moon. He motioned me to him and I walked to his seat at the panel. He opened a kind of drawer and handed me a metallic object which contained colored filters.

"Hold these over your camera lens when you take pictures," he instructed.

I remembered I should be taking pictures and hurriedly grabbed my camera, began snapping through a porthole. I got an especially good picture showing cloud formations and the atmosphere around the moon; but as we came closer to the surface I noticed the photographs did not come out well. I could see we were approaching a huge dome-shaped building, about 150 feet in diameter and perhaps 50 feet high. I could see colored lights flashing inside it, shining through the translucent material of which it was constructed. As we rounded the building and prepared to land, I noticed a base or pedestal of solid white material on which the dome structure rested.

I became more aware of the simple beauty of the lovely, iridescent, pearl-like structure. As we landed, I saw we were gliding across a flat copper-colored road toward a huge opening in the base of the structure.

The door of our craft opened and we stepped out on a ramp which led down to the floor of the building, which appeared to be a huge aerodrome. Moving ramps led up to other floors, and I guessed there must be at least two levels above the ground floor.

Next we were led to a huge lounge, with potted plants and flowers lining the walls and attractively arranged near seating units. Sculptured bas-reliefs decorated the walls.

Attractive ladies in flowing pastel gowns came toward us smiling, offered us refreshments. We took the drinks and sat on a long circular couch in front of a view screen.

Several screens, without sound, were on at the same time; if one wished to listen to any one of them, he had but to push an appropriate button. The screens seemed to be transmitting regular programs from different planets. Some of them were educational, while others appeared to be designed purely for entertainment. The ladies explained we were waiting for our guides, who soon showed up.

They told us I was to split up with my companions who were going with a different group from my own. I followed along with my assigned group and came to what I guessed was an elevator shaft, as a guide pushed a button, and I assumed we were going to another level.

To my surprise the door opened upon a corridor which led to a long train-like vehicle with 10 or 15 couches with plastic domes over them. Each couch must have been 50 feet long.

The strange vehicle had no wheels, rested in suspension about a foot above the copper highway which ribboned through the terrain and disappeared from view. We boarded the "train" and soon were gliding noiselessly above the highway. As we traveled, we could see all around and above us.

If I write another book, perhaps I can at that time take enough space to describe my visit in some detail; but it would take hundreds of pages to do it justice. Instead, I shall review it quite briefly.

We seemed to be on a tour. First, we were taken from building to building, then apparently out of the city. We passed mountains, went through valleys, visited underground installations; every few seconds people of our party would give out with "Oh's" and "Ah's," as some new breathtaking sight appeared.

Some of the terrain, in one section of the moon near the so-called "dark-side" reminded me of Flagstaff, Ariz., while other desert sections made me think of Nevada. Huge cliffs and mountains made our own look like ant hills. One particular desert locale brought to mind "The Valley Of Fire" in Nevada. There we stopped long enough for our guide to open the door and permit us to stick our heads out for a brief moment, which was all one could take, for it was terribly hot outside—like a blast furnace. I was certain no one could have lived outside very long and was glad he had shut the door.

At that moment a huge bullet-shaped object, broken and protruding from the sand where it had crashed, came into view, giving silent testimony to man's pitiful attempts at getting into space by brute force. Our guide confirmed that it represented a brave attempt by some unknown planet, and at the same time spoke with great respect for whom he termed "intrepid men from a distant world."

Apparently the rocket was the second stage of a much larger craft. Watching the tragic scene as it quickly flashed by, I assumed that the end of the craft, consisting of four spheres which looked as if they were supposed to revolve, had contained the men, and should have separated from the second stage in order to effect a landing. Something had gone wrong as it had remained attached.

Still not naming the planet of origin our guide added the rocket had crashed there in that blast furnace of a desert in the Earth year 1944.

Finally we came to another large dome-shaped building, where we halted and our guide told us we could get out on the moon's surface where we could breathe the air with little or no difficulty. That pleased the group, for we were eager to stretch our legs.

My first impression was that I was in the desert. The air was warm and dry. I could see little wind funnels forming on the ground, drawing up dust particles like tiny whirlwinds. I looked up at the sky. It was a yellowish color. When looking I had the queer impression that if I walked some distance I would fall off, since the horizon seem foreshortened.

In the distance we could see the jagged edges of high mountains etched black against the saffron-colored sky. The ground beneath our feet was like yellowish-white powdery sand, with stones and boulders and some minute plant life showing here and there as we looked around us.

Along with its weird beauty, the landscape of that side of the moon had an air of desolateness difficult to describe. I remembered wishing the rocket we had seen had tried landing on the other side of the moon where the crew would probably have had a better chance to survive.

Once again we were separated into smaller groups, according to language, and each given a guide who spoke its particular language. Along with ordinary folk, scientists, geologists, electronic engineers, rocket experts (one of whom I knew personally), astronomers (I also knew one of them) and other learned people made up our group. In the other groups I had spotted hundreds of Russians, Japanese, Germans and other people from other nations. Despite the language barrier, however, it seemed all the people in the tour had a kind of common bond of understanding and brotherhood; warm smiles and hearty handshakes abounded when there was no vocal way of communication.

Since I was an ordinary lay observer, I was not shown many of the things the technicians were allowed to inspect;

but I probably would not have understood the concepts anyhow. All of us were shown musical instruments, samples of art and architecture, and other interesting things. In fact one building was like an interplanetary world's fair, with each planet represented by some sort of contribution in art, technology and so on.

They also showed us their advanced horticultural operations, and in one place I saw flowers and plants growing in long vats of jelly-like substance. We were shown how clothing was cleaned by a kind of high-frequency sonics, and passed, in one building, trays of exquisitely cut gems which we were permitted to handle and inspect. The multitude of sights we witnessed was enough to stagger the imagination. Our wonder probably could be compared to that of an aborigine from Australia on a sight-seeing tour of New York City for the first time.

After four days of this lunar junket, we were finally treated to a huge dinner by our hosts, with such a spirit of happiness and good will permeating all of us it made me wonder if what I was seeing and hearing were not just all a beautiful dream.

But I had been able to take photographs to prove my trip, though only of the dome-shaped buildings, the craft and some mountains (for some reason I was never allowed to take photographs of surface detail, people, their mechanical installations and the like).

The dinner signaled our departure and, once again in the ship, it seemed our reprocessing consumed very little time. Before we knew it we were back on Earth, disembarking at the same field from which we had left.

As I drove off in my car, I wondered if the storm which we had seen brewing over the South Pacific on the view screen would hit soon or be dissipated before it reached the lower atmosphere.

BOOK TWO

LANDING!

PART 1

QUESTIONS AND ANSWERS ABOUT FLYING SAUCERS

Because of the many letters asking for certain information, and the question and answer period after each of my lectures, I have thought it helpful to collect some of the most-asked questions to present, with my answers, to the readers:

Q. What is a flying saucer?
A. An interplanetary space craft.
Q. Where do they come from?
A. From other planets in this system, such as Mars, Venus, Saturn; also from planets outside this solar system.*
Q. Who has seen them?
A. Thousands of people all over the world.
Q. What do they look like?
A. Saucer-shaped (as the name implies), disc-shaped, bell-shaped and so on. They may often appear to take on different shapes and colors due to the magnetic fields surrounding them.
Q. Where do we see them?
A. In flight, in skies all over the world. They land only in secluded areas where they may contact people without attracting attention.
Q. How fast do they fly?
A. In the earth's atmosphere they travel in excess of 20,000 miles per hour; outside the earth's atmosphere they can exceed the speed of light.
Q. Has anyone seen them land?
A. Yes, many people have seen them land—such as myself, and countless others who have not told of their experiences.
Q. Do they have people in them?
A. Yes, physical beings like ourselves operate the craft.
Q. Do the people get out?

* There are also space craft, though of inferior design, which are built by people of this planet. These people are in communication and in service with people from other planets. They are people who possess a high spiritual understanding and have reached an awareness of natural law; therefore they have been entrusted with information enabling them to construct such craft.

A. Yes, when they are to make contact or gather information.

Q. What do the people look like?

A. They are humans and look just like we do, excepting their manner of dress. They have solid physical bodies.

Q. How many people are aboard a flying saucer?

A. I have never seen more than six in one craft; however, they can travel in units of 3-6-9, or 4-8-12, depending on the planet from which they originate, or the polarized balance of the people in connection with the mechanics of the craft traveling through space.

Q. Do the people say anything or communicate in any way?

A. They communicate telepathically and orally with whomever they may contact.

Q. What do they say?

A. They say they come in love and compassion for us, their brothers, to help us to help ourselves to reach a higher understanding of life and its meaning.

Q. What language do they speak?

A. They have their own language, unintelligible to use because of the higher frequency and different harmonics in the tonal scale; however they can speak any language on Earth after a short period of study aided by electronic instruments.

Q. Do they indicate where they come from?

A. Yes, usually. The ones who have contacted me have come from Mars, Saturn, Venus and probably Jupiter.

Q. Do they seem to be peacefully disposed toward us?

A. They say no man can leave his planet with the purpose of conquering or controlling another world. They are not hostile. They come in love and service to the Infinite Father.

Q. Did anyone ever see a craft take off?

A. Yes, hundreds of people have seen them take off and have ridden in them.

Q. Do the craft make any noise?

A. None audible to our physical ears.

Q. Are there any pictures of craft taking off?

A. Yes. I have color movie films of craft taking off and landing, people getting out and stepping into the craft. I have Polaroid shots of the same, which will be shown in the future.

Q. Is the viewer of a craft affected in any way—emotionally, physically or mentally?

A. What happens we ourselves may unwittingly cause by fear or panic. Some cases have been reported where an individual got too close to the craft while the power was still on.

Q. If there are such craft in our skies, why are they not a more common sight, such as our regular aircraft?

A. First, they are not our regular aircraft; second, they are considered alien to our skies; thirdly, it has to be a slow process in reaching the people because of the hostile nature of this planet.

Q. Why don't they make proper visits through channels of government, or mass meetings and landings?

A. Mass landings, great displays, and the like would only cause confusion. The military would be involved immediately; the governments of the world would be in turmoil, each seeking its own advantage. There would be hysteria and, possibly, panic. And so, in the interest of humanity, the space people approach us cautiously. Incidentally, space visitors do not have identification papers and passports. If we now investigate and counter-investigate every suspected alien who might be suspected of being a foreign agent, what would we do if confronted with people who are entirely new, strange, and alien to our planet? There would be endless investigations and controversy, and the work and message the space people have come to deliver would be snowed under by red tape. I doubt whether they have the time or inclination to play this silly Earth game of intrigue and counter-espionage. So they come directly to the people, by contacting their own; and the people will learn of them over a period of time—gradually, without fear, panic or censorship. Every great movement has always started with the people, and that is where the space story belongs: WITH THE PEOPLE.

Q. If they can speak our language, why don't they come among us and announce themselves, as we would do if we went to another planet?

A. They have tried, but people would not believe them—mainly because they look and act like us, and are not monsters with six or eight arms.

Q. How long have they been coming into our atmosphere?

A. For thousands and thousands of years.

Q. Why don't they tell us about the wonderful experience of space flight?

A. They have, through certain people, and these certain

people are willing to listen and believe whether or not they see; and also some of the information divulged must be kept secret.

Q. What is the place of the space people in the cosmos—do they come from different dimensions?

A. They are spiritual beings, like ourselves, using physical forms adapted to their own planets. There is time around their planets within their atmosphere just as on Earth; but there is no time in space.

Q. Is this their first or early beginnings of flight into space?

A. No, they have been traveling in space for thousands of years.

Q. How do they regard us?

A. As brothers. They love us.

Q. Why do they come here—what is their purpose?

A. To try to awaken within us a yearning for higher understanding so we can help ourselves in preventing any further destruction of our planet, which could conceivably have a bad effect in our solar system. It is about time we grew up as a humanity.

Q. From what planets are they coming?

A. From Mars, Venus, Saturn, Jupiter, and some planets outside our own solar system.

Q. Are any craft coming from beyond our solar system?

A. Yes. And some mother ships have come from distant galaxies.

Q. What are the main different types of craft?

A. *Discs:* are remotely controlled objects sent out from the masterships, to record thoughts, emotions, feelings and other conditions of the people in the area; also to detect hostility before landing. Some of these recordings are used for future contacts with the individuals concerned. These discs range in size from eight inches to more than eight feet in diameter.

Bell-shaped Saturnian craft: about 50 feet or so in diameter, 18 to 20 feet high; metallic, grey, somewhat flatter in appearance than a Venusian scout ship.

Mother ship or carrier: eliptical, cigar-shaped, or egg-shaped. They have been reported up to 3,500 feet in length; however, there is no limit to their size.

Green fire balls: are means used by space people to protect us from the effects of the atomic and hydrogen explosions in our atmosphere.

Q. What are other planets like, for example, Venus or Mars?

A. Venus is a planet slightly smaller than our earth. It is in the stage that the earth was in many thousands of years ago: young and healthy, with beautiful foliage, streams, forests, large bodies of water, mountains, hills. As a matter of fact, there are some places in California today that resemble Venus. It is beautiful and verdant and a veritable paradise. There are also places in South America similar to places I have seen on Venus. They plan to keep their planet young, beautiful and healthy. Their atmosphere is very similar to ours, but the sun cannot penetrate it with destructive rays. The people are predominantly light-skinned and fair.

Q. Do they have governments, cities, country places, farms, gardens, factories, schools, etc.?

A. They do not have authorities or government officials of any kind. They live in peace and harmony and everyone knows what his or her particular talent is so that they work at that particular job—and they love their work.

Factories: There are buildings where work is done, or where the craft are built; but the buildings are beautiful places and not like our factories at all. They receive no coin in exchange for work. Instead they exchange talents, and everything is shared to the extent of their talents and desires and no one wants for anything. We work because we have to work. They work in service of their Infinite Father.

Farms: They grow fruit and vegetables and flowers. They do not raise meat-producing animals since they do not eat meat. Animals roam free and complete their cycle of life naturally. They are not over-bred and over-produced for food.

Schools: There are schools of wisdom where children or adults can attend. Most knowledge is inherent in the children, since they are born with the knowledge of the past. Their education is their past learning, which they apply in their present life to gain wisdom for future use.

Cities: They live in small communities, built in the forests and close to natural surroundings. They do not denude the land of all trees and shrubs and then build boxes. Their communities are kept small, usually contain no more than a few thousand people. They are spread out and decentralized.

Q. If a well-known people have been contacted, why don't they tell about their contacts?

A. Government officials in particular refuse to tell

because it would upset our economy. The knowledge they have gained depicts an entirely different way of life. It is living under God's law rather than man's law. Most mechanical energy sources would become obsolete.

Q. Have all countries been contacted?

A. People in every country of this world have been contacted.

Q. Why do they contact only certain people?

A. Certain people are born with an awareness of truth within themselves, or they are reborns from another planet, in which case their own are contacting them and awakening within them that one small spark of truth so that they "become the flame of truth." These people must have the courage of their own convictions and the ability to "take it," for they will suffer ridicule and attacks.

Q. If they are coming here to help us, why are they concealing their identity?

A. They are not particularly trying to conceal their identity, and to those they have contacted they have revealed themselves.

Q. Are space people living here on the earth among us?

A. Yes, thousands of people from other planets are living among us. Some are rebirths, some have come directly from their home planets in spacecraft. They may live next door to you. One of them may be your co-worker, the person who serves you in a store or restaurant. They have one identifying trait: love of fellowman.

Q. Why don't they tell us how to build a craft?

A. Because it is like giving a child a firecracker, an automobile or gun. We cannot live with ourselves, let alone trying to live with people of other planets. We would use this power to conquer. On other planets there are no wars—and they would like to keep it that way.

Q. If, as some reports indicate, some of our airline pilots are seeing flying saucers and are said to be suffering from hallucinations or seeing weather balloons, why are they not fired since the safety of our flights depend upon the pilots?

A. Because authorities know the pilots are NOT suffering from hallucinations. They know the pilots are telling the truth, and saucers are being seen by too many airmen to adequately squelch the story.

Q. What is the average life span on other planets?

A. Eight hundred years.

Q. Have the space people brought us any pictures or films of their home planets?

A. Yes. In the future pictures taken on other planets will be shown here. These will include scenes of the planet, people, animals, etc.

Q. What kind of clothing do the space people wear?

A. On Venus and some planets, the women wear knee-length, billowing, tunic-like gowns of pastel colors. Some have no sleeves; yet some have long, full sleeves. The waist is sometimes held by a jeweled belt. The women do not wear girdles, bandeaux, or any tight undergarments. The clothes are comfortable, airy, loose and beautiful, enhancing the contours of the female form.

The men wear ski-type trousers, translucent and soft, something like nylon. The clothing both for men and women adjusts to the bodily heat so that it can keep them cool or warm as the temperature varies.

A sandal-type foot gear is worn by both men and women.

Q. Do they have families, children? What is their social setup?

A. Two beings when perfectly mated stay together as long as their desire and mutual progress continue, sometimes for many of our lifetimes. They have children, and the children are loved by all. However the children mature at a very early age.

Their social setup is communal. They share the goods of life with one another. Yet, if it is their desire, they can have isolation and privacy at any time they wish.

Q. At what age do children reach maturity on other planets, such as Venus?

A. They mature in three to five years. A child on Venus shortly after birth is already equal to an Earth child of seven.

Q. Do they nurse their children?

A. Yes, the children are breast-fed a few months, then are weaned on natural foods, such as fruit and vegetable pulp. They are not given animal milk.

Q. Do the children go to school?

A. They have communal type schools or places where they are briefed on their own spiritual development. Most of their knowledge is within themselves and in such schools it is encouraged to develop.

Q. Do people work?

A. There is no work as we know it. They have advanced mechanics and apparatus that do the work quickly and efficiently. All services are voluntary and rendered with love. All products are shared. They do have buildings where people go to perform services, where the various conveniences of life are made.

Q. What is their religion? Do they believe in God? Do they believe in Jesus?

A. Their religion, or, more properly, way of life, is serving their Infinite Father, and attaining more knowledge, so they can serve their Creator to a higher degree.

Jesus was one of them in the highest degree of development.

Q. What are their homes like?

A. On Venus, the buildings are dome-shaped and semi-translucent to permit light and color to enter. Some of the buildings resemble our own modern organic architecture.

Q. What is the weather like on other planets? Do they have seasons?

A. The seasons are not as drastic. For instance, on Venus, as it is on Earth, they do have seasons, although, the temperature is generally the same all year round—like perennial spring; however, in some sections they do have a change in season and temperature and experience winters.

Q. Can we go to the other planets and be accepted and live as they do?

A. Generally, no. We might, due to the difference in development and vibration, atmospheric pressures, etc., suffer a nervous breakdown. Some people from Earth, however, have gone physically to other planets aboard a craft and stayed there (not wishing to return); others have returned after a learning period or visit, to help their Earth brothers. Those who do return usually remain silent as to their experience lest they be confined to a mental institution or suffer ridicule.

Q. Can a disc-type craft return to Venus without having to depend upon a mother ship for transport back?

A. Yes, the craft has the ability to return to Venus on its own power without transport by a mother ship.

Q. What is the purpose of the mother ship or carrier?

A. The space carrier is used outside this system for extended trips into outer space. It transports cargo and equipment, along with many people.

Q. Can space people take goods or plants from here back to their own planets?

A. Yes. Many craft are sent here for specific botanical studies, and take many of our plants back with them for study and transplanting.

Q. Can we transplant plants from their planets to Earth?

A. Yes. Plants can adapt themselves to the conditions of varying frequencies in time.

Q. Have some of our terrestrial plants been brought here from other planets?

A. Yes. Some of our plants have been brought here from space.

Q. Are the planets known by names different from the ones we call them?

A. Yes. In some cases. In other cases planets are designated by symbols. Our Earth, for example, has a specific symbol.

Q. Do people from other planets have hair on their bodies as we do?

A. They are not as hairy. In some cases, however, when a person from Venus comes here, he will, after a short while, grow hair much faster than the ordinary Earth person. When they return to their own planets they lose the hairy condition.

Q. Does the frequency of a planet affect the mental development of its people?

A. Yes.

Q. Are men and women equal in social level on other planets?

A. Yes. In fact, from a physical standpoint it is more pleasurable to be a woman on other planets. Childbirth, for instance, is a thrilling and pleasurable event, not associated with pain or discomfort.

Q. What is the difference between a reborn and a reincarnated being?

A. *A reborn* (or rebirth) is a being who has volunteered to come to this planet from a higher planet or dimension, on a mission to teach his or her brothers and sisters and assist them in helping themselves gain more insight and understanding of the Universal laws of the Creator.

A reincarnated being is a direct rebirth from this planet due to a compensative hereditary expression of cause and effect in the physical illusion, or in other words, karma.

Reborns, however, teach that one can live above karma, and that understanding and awareness is a higher law than cause and effect.

Q. Is it difficult for a reborn to understand the reasoning of earth people? Is their reasoning or understanding different?

A. Reborns are usually masters or near-masters; therefore, it is not difficult for them to understand the Earth reasoning or its people; however, it is, at times, difficult to live among them.

Q. What is the difference between a master and a near-master?

A. A *master* can do anything, but does not.

A *near-master* (such as an adept) can do almost everything, and does. There is that small percentage of ego still operating that prefers to demonstrate through the individual ego, rather than operate God's law through many forms and people.

A *master* appears confused, but is not, among those who are confused, just to be one of them.

A *near-master* appears calm and in control before the confused, and prefers to remain aloof from humanity.

A *master* can leave (mentally and etherically) a group of people engaged in conversation at any time he wishes, without the people being aware of his spiritual absence, or partial absence, since a master can be several places at one time.

A *near-master* can leave, mentally *or* physically, and appears before those people then, as a master.

A *master* directs indirectly the laws of God through the confused people around him, because when they act on what they term as their own thoughts, these actions are recorded on their own subconscious and they then learn through their own mental processing what the master has known for centuries.

A *near-master* directs throughout what *he* knows as God's laws and the beings in the physical illusion do what they think *he* knows is right and do not learn as much.

There are four real masters in the world today. One is in the United States, one is in India, one is in Australia, and one is in South America.

The near-masters do not know of the masters, but the masters know of the near-masters, and of the reborns, who do not know of themselves as near-masters and/or reborns. This realization of one's near-mastership usually comes somewhere between the ages of 30 to 40.

Q. Are the masters celibates?

A. No. God is married to the Infinite Universe. All masters are married. It takes a master to get married and perform physically that which is taken for granted that a master would not do. Sometimes it is a matter of choice. The sexual expression is one of the highest expressions of God, in that it takes in the sense of touch. When we speak of sex, we interpret it to mean physical union expressed through love and understanding—not sex as expressed for and by itself. Sex expressed purely on the physical plane for itself and in improper balance with another being is adulterous. Two beings, perfectly combined in all areas, that is, in the spiritual, mental, emotional, and physical, should be joined as one. Wedded incompatibility is the true adultery.

Q. What is a flash-back?

A. A mental picture of an experience, or a feeling of remembering a place or lifetime on this planet or another planet. Only brains and mental makeup of reborns can tune in to flash-backs. Certain granules in their brain cells, which by previous development, will respond to flash-backs.

Q. Why don't we remember past lives?

A. The memory of past lives is trained out of us from childhood onward. Children's fantasies, imagination and play are sometimes really flash-backs and/or regressive memory.

Q. How are the visitors trying to raise the mass consciousness of the people?

A. By various methods such as:

a. Dissemination of saucer research data.

b. Stories of contacts with their own.

c. Their signs in the sky.

d. Mechanically by means of mental capsulation, and machines.

Mental capsulation can be projected by sound, color, vibration. A high-frequency sound can be a mental capsulation; a song or a selection of music can be a mental capsulation.

Music in the form of a mental capsulation helps "push certain buttons" in the mind and releases something which is there. However, because of the way we eat, think, and act on this third-dimensional planet, this knowingness is held dormant, sometimes through several lifetimes. Nevertheless, God's laws shall prevail, and those who are due this release toward an awareness will have it when the time in the cycle of expression of this planet is ripe.

The machines which send out super-sonic high frequency sounds use a man's body as a terminal in conjunction with the mental capsulation. There are three terminal bodies in each state. The machines now operate on a silent carrier wave.

Q. What is the Will of the Father and how can it be developed?

A. The will of the Infinite Father is to EXPRESS in ALL dimensions, the love of the Infinite Father in ALL FORMS, COLORS, SOUNDS, TASTES, AND EXPRESSIONS.

Q. Will there ever be an interchange of peoples, ideas, and cultures between the planets in our solar system, so that there will be brotherhood throughout our solar system?

A. This is inevitable. You can delay God's plan, but never stop it. Interplanetary brotherhood for earth's people is dependent upon the degree of decline of hostility and the degree of increase toward tolerance, love, and good will toward our fellow men.

PART 2

A PROFOUND SPACE TEACHER

In late August, 1956, I had what was perhaps the most memorable meeting of all when I was privileged to meet one of the most advanced beings from another planet.

I went to Field Location No. 1 just at sunset, and the craft, coming in from the west, looked like a huge sun descending from the sky.

I remembered hearing the accounts told of the Miracle of Fatima. Witnesses had described a huge, fiery ball, which they thought was the sun, moving in the sky, as if headed for earth. I wondered if there had been some connection with the space people.

This craft appeared much different than the usual type. In fact I smiled to myself as I noted it looked like an inverted translucent china cup (however, without a handle!) set into a saucer. This ship was larger than the others and had portholes in series of four. Like our own aircraft, the vehicles used by the space people are of many different types and designs.

When it landed I walked to within a hundred feet of it when two men stepped out of it and waited, one on each side of the door. I sensed something important and unusual was about to take place. I walked closer, then stopped, waiting for a feeling or sense which would indicate it was all right to proceed further.

Then a magnificent sight appeared in the doorway. A tall handsome man with long blond hair over his shoulders stood towering at the entrance. He looked directly at me as he stepped out of the ship.

Then he came toward me. But he seemed to float or glide rather than walk. His body appeared to be weightless. When he was about 25 feet from me, he stopped and I continued walking toward him. When I was about ten feet from this magnificent creature, he raised his arms, indicating I was to stop.

The sight of the man thrilled me. He was dressed in a radiant white ski-type uniform, girdled about the waist with a white belt. The garment was tight at the ankles and wrists, yet the sleeves were full and loose; the neck-line was high, similar to that of a turtle-neck sweater. Over the uniform he wore a light blue, fluorescent-like cape, fastened at the left shoulder with a gold pin in the shape of a wheel.

A slight, enigmatic smile played on his lips as he looked at me, while the expression in his eyes was one of pure love and understanding. I felt like a small child, humble and loving, and I wanted to embrace him as a long lost friend and remembered loved one.

I believe I had the same feelings I would have experienced had the Master Jesus just stepped out of the craft. I studied his face—it was all compassion and love. His skin was fine and white, his nose straight. His eyes were the color of goldenrod when it is ripe; yellow as the oblique rays of the setting sun was the equally radiant light of his eyes. His fingers were long and tapering. Truly, he was beautiful as a woman, but with the bearing and masculinity of a strong, virile man.

The moment he raised his arm, he began to communicate with me telepathically, and in five minutes of such transmission he gave me more knowledge than one could absorb in a week or more of solid study and oral transmission. One might compare it with the same situation ascribed to a drowning person, whereby an entire life is said to flash before him in a few brief seconds. In the same manner, certain knowledge was transmitted to me in a few minutes which would pertain to life and the future.

This transmission took the form of pictures. Mental words would have taken too long, and, as the Chinese say, "One picture is worth a thousand words." It could be described as a form of telepathic-television type of communication. I received the mental pictures as rapidly as a moving picture and occasionally heard the sound of his rich, deep voice blending into the pictures. Some of the concepts I received were too technical and advanced for me to understand. But at the same time he assured me I would retain the knowledge as a kind of mental capsulation, which would come out in small degrees over a period of time.

I would be giving this information to people when I spoke to them in small and large groups. It would not be necessary for me to prepare a speech; the words would come to me as I recalled the information now being given me. When people asked questions (at that time I didn't know how many there would be!) I would have the answers. Somewhere in the depths of my mind the knowledge was implanted and I had but to call on my subconscious to bring it forth.

I shall never forget this awe-inspiring experience: it is too strongly impressed in my soul by the love, compassion and wisdom of this profound teacher. That wonderful being unlocked the door to my own soul, and in a few brief moments planted the seeds of infinite knowledge in my subconscious, which over a period of months and years gradually seeps out to my conscious expression. It reminds me of the ministry of Jesus. His entire message of love and good will toward mankind was a mental capsulation to the people of his time; and it is only now that the truth of his being is gradually coming out and the understanding of his great love is emerging from the quicksand of misunderstanding, at last into the sunlight of our conscious acceptance of the Infinite Father.

Following is only a poor attempt to set down some of the words this teacher spoke or conveyed to me, and it is with humbleness and gratitude that I present them to my brothers on Earth.

Truth never was, never will be a theory, nor contemplative arrangement of philosophy of men, nor intellectual insight in the minds of men. Truth IS. Great men on your planet think of truth in a different light, as compared with your brothers from "Space." For these men, Truth communicates with reality, which is the real nature of that which he perceives with his six senses. Man continually seeks his Source, the Supreme Consciousness—and those great men of your holy writings—who touched upon this Source, discovered a divine plan for all mankind, one rooted in love; for the Supreme Consciousness is love. We are dedicated emissaries of this divine plan, to your planet, to those with an evolved insight, for those are the ones who will receive us.

You must act with realism in the illusion as you walk in the light of the Infinite Mind among your people. You are an illusion or projection in a given dimension of your real self. You call it third-dimensional, but this is not accurate, because you could not see, hear, taste, smell or touch unless you were, in effect, a reality, an expression in a projected form of fourth dimension.

The very fact that you can think makes you a fourth dimensional being. Inanimate objects, such as tables, automobiles, houses, are in reality third-dimensional. Thinking is your sixth sense and you do not have to perceive any of the five known senses to think. However, a movie, or televised picture is just as real, with the exception that the projected movie or television are merely a reflection, very limited in that it does not think or control itself.

It has been stated by some of your great thinkers that time itself is the fourth dimension. Time is a condition of the fourth dimension, because you cannot have time without motion; nor motion without time—and neither without thought. You, as an expressing, thinking being, are composed of both motion and time, making you a fourth-dimensional thinking being.

Nothing we see with our physical eyes is Truth, but simply a reality in the dimension of a reflection, or an effect, secondary in nature related to a Cause from a primary Source. However, the mind, be it known, still thinks after the so-called "death," which in reality does not exist. In Truth, neither life nor death exists. Truth IS.

Truth never changes. Only reality, in the form of matter, energy, and time changes. Your fourth-dimensional body is expressing as an instrument of your infinite spirit, through soul and mind, with the brain acting as an instrument of your mind, as a radio or computor receives what you call electricity to activate its motions and express with music, voice or answers to problems. The electricity still exists after the radio is destroyed, although both are inferior by far to the mind and brain.

Some of your metaphysicians say that you are the sum total of what you have been in past expressions. This is highly inaccurate. You can only be the sum total of past experiences if you are fully aware of those experiences and lessons. Forgotten lessons cannot be added to the total until one is made aware of them. You are never the sum total in any given dimension.

Many of your population have voluntarily reincarnated here on Earth from other planets of your own solar system, to help in a plan which is universal in scope. They, on their previous planets, have expressed and experienced a much higher understanding of the Infinite Father's universal laws. We have only begun to contact them and release them from a memory block, due to the lower frequency of your planet Earth. Nevertheless, we dare not interfere in the form of force or control in any way. This would not conform to the Father's laws to which we adhere. We are for progress, fulfilling to the best of our capacities, the Will of the Infinite Father. To act against these laws, we realize, would turn them against us.

But, this is not the reason we do not act against them. To act against them would deter progress of expression, the very progress to which we have dedicated ourselves. Many of you, the people of Earth, those reincarnated volunteers, have had quick flashes of a truth communicated to your minds originating from your infinite self, but these thoughts have been discarded as imagination, or hallucinations and dispelled from your consciousness. It is easier to do this than listen and act upon truth. It is easier to conform to the distorted ways of your world than to speak your mind and lose so-called friends, prestige, money, power, and false security. This is the difference between the men you have named Jesus, Moses, Buddha, Confucius, and many others, as compared to kings, generals, tyrants and dictators who are all in reality strategic murderers of men.

Your scriptures tell of the evil of murder and persecution as evidenced by Moses and Jesus. This being against the laws of the "Being" you call God has not changed the thinking of the people of earth. You still look up to, fear, or conform to men who have reached a high place of authority or power over their brothers. Do you think this God sanctions force, persecution, murder or evil in any form? Certainly not. Those who do, worship a false God, to suit their own ends.

Soon, your closest "friends" will shun you in the near future because of their distorted interpretations of events that will take place, but this is necessary, in the sense that it becomes part of a large screening process, all closely monitored by us, who are orbiting your planet in answer to the many prayers for help from your population of many beliefs, colors, creeds, races. We would not come to the aid of Earth unless there was a call for help on a mass scale. In the year of your time 1945 we concentrated a large group of servants of our Infinite Father, which you have called "angels" in your scriptures, to aid you with the help of our machines, which project thought impulses in the direction of a specified area of chaos. In our space craft outside of your atmosphere our instruments received, telescreened and recorded the prayers of millions of souls, desperately asking their particular "God" to help them in their suffering.

There was a large concentration of these thought impulses in the form of prayer in the area you call Japan where your fissional explosions took place and destroyed thousands of bodies in a horrible way. In another area, your United States, thousands prayed to another God to help their armies to make haste in killing the Germans or Japanese enemy and end the war so their loved ones could come back to them.

We noticed and recorded your distorted concepts of what God is with great sorrow and concern for your ignorance. On our view screens thought patterns of God took on many forms and shapes, but most were men, some with beards, some tall, some short, and of course, some were stone or wood or metallic idols. Son, we have not seen the Supreme Intelligence in the sense that is a particular form or shape. God is not a man. To call God a man is to limit that God. God is all.

Man is limited, but God is unlimited, infinite, expressing in all men, all forms. Men are student gods, going through a school of expression on this planet and many others, seeking knowledge and wisdom so that he may serve his brothers and the Infinite Father of creation. Man continually progresses up the ladder toward perfection, and though a rung may break under the weight of his many errors, still his goal is to reach the top and one-ness with the Infinite Father. His soul records every experience, thoughts, mistakes from the time it becomes a soul. A man's soul, as with the lower forms of life, such as dogs, cats, cows, horses, etc., is the sum total of a process of evolvement of consciousness. Everything created has a consciousness, and the conciousness evolves to the soul point, where it expresses in higher forms, man being the highest. This doen't necessarily mean one evolves from the other, but there are many cycles of evolution. Man's infinite spirit is perfect, and uses an infinite body in a given dimension possessed of a soul, the recorder, closely related to the mind, which does the thinking, using the finite brain or instrument of perception in conjunction with the "five" senses.

What you call "reincarnation" is preceded by a transition or change, called "death," but this is not the end of the consciousness, but a continuation of a type of experience leaving certain "senses" behind on the physical plane and becoming aware of a more enlightened consciousness, and you discover that you are still thinking without the instrumentality of the brain. Son, I say to you that there is no such thing as death in any form, for if there were, you would not be here now, for you always were, always will be—you are, as it also is with the universe, as it is with our Infinite Father of Creation.

Your scientists are necessary in furthering progress of all aspects of life on your planet. When "science" comes about, man begins to find the answers to many phenomena observed, but not understood. However, now science is actually limiting itself, and the progress of your populace, by sanctioning that which it is able to prove only through objective reality, rather than subjective reality as such, related to truth. A scientist observes visually in the course of a particular experiment an occurrence which did take place before his eyes, but, because he did not know how or why it happened, he rejects it, and it is not "scientific" fact until he find the answer within the scope of accepted scientific theory. He probably never will find all the answers using science in its present form, so there must come from the scientists of your day a new science—something beyond science—before progress will continue in this realm.

This master of a man transmitted to me a picture of truth, but—I realize my inadequacy in conveying a spiritual perception inexpressible in words. Each individual would probably interpret his message in a different way. This is the sad part of their contacting us on this earthy plane of expression, where we are very limited in interpreting mental impressions and meanings into words. The very fact of this type of relay from one medium (telepathic) to another (memory recall) is comparable to subjective reality as reflected in the form of objective reality. Thought expression is restricted to language. Telepathy is the clearest and the most direct way of communication.

I have tried to recapture some of the thought impressions I received from this great teacher. The meeting with this individual has given me the encouragement and incentive to continue with my work.

PART 3

THE EXPERIMENTS

Although I would like to take personal credit for it, I believe the space people were responsible for my performing an amazing experiment early in 1956. Of course inventiveness runs in my family. Both my grandfather and great-grandfather were artists and inventors. I like to feel that I am similar to them in that respect.

Although I had not thought too much about inventing things, early in life I had wanted to become an artist; but it was impossible for me to obtain the necessary training. Perhaps I sublimated this urge in the more prosaic occupation of sign painting.

Often in the sign shop, after working hours, however, I would putter around, trying to make new things.

On one such evening I had just finished a sign and was cleaning the brushes prior to going home when the telephone rang. I said, "Hello!" into it, but there was no response; so I said, "Hello!" several times. Still no reply, so I put it down. Immediately, however, I had the urge to lift the receiver and listen again. I did so, and there was no dial tone; the circuit was still disconnected.

I listened for a few moments and was about to replace the receiver when I heard a high-pitched buzzing sound coming from it. I put it back to my ear, but after a few seconds it stopped. Then I put the telephone down and forgot about it.

As I was preparing to lock up I received a sudden and strong impulse to remain in the shop. Then almost mechanically I began to move about, pick up pieces of wood, other materials, and set about constructing something—just what I didn't know. Whatever I was doing, I seemed to be controlled and directed, working as if I had a set of blueprints right in front of me and knew exactly every move to make.

I set the materials I had assembled on the bench. Using a 24-inch plywood base, I drilled a hole halfway through the center. In the hole I placed securely the carbon rod of a broken flashlight cell. Next I obtained a large nail of like diameter and wound the nail with 50 turns of very fine copper wire. When I had slid the finished coil off the nail, the unwound ends of wire remaining measured about eight or 10 inches. I soldered two extra connections: one from the top and one from the bottom; two of the ends were connected to a small pen flashlight battery. I stapled some of the wires to the plywood base to hold the assembly in position.

Next I took the small brass cap off the larger flashlight battery cell and glued it in the exact center of a 10-inch aluminum disc, which had been put on the jigsaw and placed in a balanced position directly on top of the carbon rod. I do remember definitely that four ends of the wire pointed at the edges of the disc directly opposite each other in the form of a cross, and approximately one-eighth of an inch from the edge of the disc. All of the equipment was set up in accordance with an imaginary shape of a pyramid.

Then I made two connections at the bottom of the apparatus, which apparently completed the circuit—for the disc immediately was enveloped by a bluish, spinning light! Then, to my utter amazement, it rose from the platform, crashed through the 12-foot ceiling, which was made of aluminum paper, apparently bounced off the peak roof and then came down again through the same hole through which it had made its ascent!

It crashed with more force than the weight of the aluminum would normally permit, almost completely destroying itself so far as reconstruction of that particular model was concerned.

I was so surprised and shocked I sat down in complete silence and thought for several minutes.

Did the electrons which came from the small battery start something in the atmosphere in the form of some type of energy which took over, once the machine had been "primed"? The small amount of power from the battery surely never could have moved the object in any way.

I was too shaken to do any more work that night, but the next morning I tried to duplicate the experiment,—without success. I just couldn't remember where I had been apparently shown to make the various connections. Later I was able to duplicate the experiment to the extent that the disc glowed; but never did it move. I am writing this with the hope that some open-minded physicist or electronics engineer will, in my brief account above, detect a key which will open the door to some type of free primary energy which the space people say is available all around us, waiting to be tapped.

Some critics, hearing of my experiments, have suggested that I photographed such models while they were in flight, and their remarks amaze me; for had I been able to invent such a model which would fly, this event in itself would be almost as startling as the spaceships they accuse me of hoaxing up.

The success of the initial experiments quite naturally put my mind on what one of my friends termed "the free energy kick," and I stepped up my experiments with magnets, which had always fascinated me and with which I had played around in little experiments in the shop.

The same friend told me one day he had invented a magnetic motor and was in the process of having it patented. I became enthusiastic, and, wishing to help him get some needed publicity, I suggested he go on the Long John Show and tell his story. I asked also that he bring a model of the motor; he agreed, and I made the necessary arrangements for his appearance on the show.

Two days before the show he telephoned me frantically, saying he would have to construct a new model, and asking if I would bring some of my magnets over to his place and help him. So we went to the laboratory of a friend of his, who is an electronics engineer. There we worked several hours, but were unsuccessful in putting together a model which would work satisfactorily.

That night when I went home I decided to rig up something quickly for the show, since I had promised by telegram to bring the man and a motor. So I stayed up all night, working out a system of off-balanced magnets, whereby the north and south poles of the magnets seeked equilibrium and came to a halt. To keep such an apparatus in motion it was necessary to upset its normal balance. This, I might add, was a different principle from the one involved in my friend's invention. I got my idea from thinking about the universe which is kept in motion because of its off-balance seeking an equilibrium (if it ever did reach an equilibrium, naturally all motion would cease).

The model I constructed for the show worked all by itself for about two minutes without my having to impart any work to the rotary, off-centered magnets. In my haste to construct in time for the show, however, I put it together with quick-drying glue, and it broke.

Since there was no time to construct a new model, I gathered up the components and rushed to the studio with them, and was able to demonstrate something of the principle involved.

As yet nobody has been able to duplicate this simple apparatus, as far as I know.

One of my latest experiments involved a thing people have called impossible—a "one-pole magnet." The minute someone says something is impossible, I am presented with a challenge, as I was in this case.

Does a magnet with only one pole refute Newton's third law of motion, which states that every force or every action produces an equal or opposite reaction? Whether or not it does, many witnesses saw me demonstrate the "one-pole magnet" during a recent lecture, sponsored by Dr. Alfred Smith of The Space Seekers Society in Philadelphia.

As with the so-called free-energy motor, a series of these magnets have turned a small direct-current generator and lighted a small flashlight bulb. In my opinion, this is an example of harnessing a free energy to impart work of its own accord, generating a secondary energy in the form of what we term electricity or electrons, which, in turn, produces light energy or heat.

Magnetic phenomena in the form of what we call attraction is not a pull, but a push from a surrounding unseen funnel of magnetic vortex enveloping this planet, which in itself represents a balance of motion in the form of imbalance which continually seeks an equilibrium—an effect presenting itself as secondary energies seen and unseen, originating from a Cause Unseen which is Infinite.

PART 4

THE MYSTERY APPEARANCE

I would like to let you in on the incredible results of one of my own experiments, and the best way to present this in detail, I believe, is to quote from one of the Long John Shows, as he is interviewing Miss Mary——:

(Jan. 11, 1957)

LJ: This morning we are talking with Mr. and Mrs. Howard Menger of High Bridge, N.J. The young couple has appeared on my show seven or eight times, and they have been telling us about their experiences of having physical contacts with people from outer space. Howard had the opportunity of riding in a saucer. We also have John Otto, Cortland Hastings, and a young lady, the sister-in-law of Howard, whom we will identify with the name "Mary." Her last name will not be used, since she is going to tell us about her own meetings with people from outer space. Mary, we have never had the opportunity of having you on the Party Line with us and we have been told by your sister, Rose Menger,* and by your brother-in-law, Howard Menger, that you have had some rather unusual experiences in the past few months. One in particular was the teleportation of a pipe. This was described to us by Howard. Now, rather than asking any questions at the moment, suppose you just take over if you will, please, and tell us in your own words what happened when Howard teleported his pipe from Pennsylvania.

MARY: I was sitting in Rose's and Howard's living room and I heard this pounding on the door, and so I went and answered it and I saw Howard there.

LJ: In front of the door?

MARY: In front of the door. I knew Howard was in the Pocono Mountain area that night. He stared—he just stared straight ahead. He did not say anything, and he handed me a pipe. I felt very strange—I-I-I just couldn't understand how he could be there. He was—He didn't come up in any car or anything. I had this verified about 10 minutes later.

LJ: What do you mean by "verified," Mary?

* The former Mrs. Menger.

MARY: He appeared at the door at 8:20 P.M. that night, and about 8:30 P.M. he called on the telephone to verify that he had been there.

LJ: Now, where did he call from, Mary?

MARY: He called from the Pocono Mountain area.

LJ: Now, how did you know he was calling from the Pocono Mountains?

MARY: Because of the call. It sounded far away. It was a long distance call.

LJ: In other words, when you pick up a long distance call you usually can tell that it is from some distance?

MARY: Yes, and also I found out later that he had been seen in this diner and he had witnesses to the effect that he was at that diner, and he had them sign their names.

LJ: Well, now, let's just take it step by step. When the operator—first of all, was this a collect call?

MARY: No, it wasn't.

LJ: In other words, Howard was paying for it on his end. Howard was calling from a diner in the Poconos. Do you have any idea how many miles away from your home in High Bridge that Howard was located at the time—approximately?

MARY: I don't know how many miles, but I would say it is about an hour and a half ride or a little more.

LJ: Now how do you judge an hour and a half as far as driving time; in other words, did you know the town he was calling from?

MARY: No, I didn't.

LJ: Mary, I am not trying to be facetious at this moment, but you certainly can't judge how far away he is by the tone of his voice; in other words, you would say that you thought he was talking from some distance because evidently his voice was rather low. Is that what you meant?

MARY: That's right.

LJ: Now, you said that he was in the Poconos?

MARY: Yes. He said he would be home in about an hour and a half.

LJ: Is that how you got the idea that it was about an hour and a half ride?

MARY: Yes.

LJ: Did he return in an hour and a half.

MARY: No, it was a little longer.

LJ: Well, maybe the roads were a little bad, and Howard, after that experience, didn't want to rush. Do you want to say something, Howard?

HM: Yes, on the way back, when I got to Washington, N.J., I went through the Easton area; that is in Pennsylvania. I wanted to call again, but something told me not to. So I went another 13 miles or so and got to the shop in Washington, N.J., and called from there instead.

LJ: Now this is not the call that she thinks was from the Poconos?

HM: Oh, no, no.

LJ: O.K. Now, Mary, to get back to you for a moment. How did Howard look to you when he came . . . when you answered the door that night? Did he appear to be in his physical body as we know it—as he looks to us tonight? Naturally, he would have an overcoat and hat on, I assume, but, in other words, did he look the same as he looks tonight?

MARY: Yes, he did, but he stared straight ahead.

LJ: What did he say to you?

MARY: He didn't say anything; he just seemed to automatically hand me the pipe.

LJ: Was this a metal pipe or a smoking pipe?

MARY: A smoking pipe, and he left it with me and then disappeared.

LJ: When you say, "disappeared"—did you still have the door open, or had you closed the door?

MARY: Yes, I still had the door open.

LJ: Did he sort of dissolve, vanish, or did he walk away?

MARY: He took a couple of steps and then he sort of disappeared.

LJ: Well, is it possible he just "disappeared" into the night? Into the darkness? I am not familiar with High Bridge, but in New York there are a lot of street lamps, and I do not think you could disappear too readily. You say he took a few steps, but, first of all, you say he was facing you, is that correct?

MARY: Yes, he was.

LJ: Did he have the pipe out as he knocked on the door, or did he have to reach into his jacket to get the pipe?

MARY: No, he had the pipe out, in his hand. He handed it to me and took a couple of steps; then I saw him disappear.

LJ: He would have to turn around first, is that correct?

MARY: I did not see him turn around. I think he took a couple of steps backward, then he just disappeared. I felt very strange. I couldn't understand it.

LJ: Well, you will have to get in line with me on that one! Do you remember what day this was—in other words, was it a Saturday, or Thursday—I am not looking for the exact date.

MARY: December 8, 1956.

HM: We did write some dates down, Long John, but, . . .

MARY: It was a Saturday night.

LJ: It doesn't matter. Now, let me continue. Do you live in the Menger's home?

MARY: No, I don't.

LJ: Do you live . . . without establishing where you live—do you live some distance away?

MARY: I live nine miles away.

LJ: When did you arrive at the Mengers' home that particular Saturday, approximately—in the afternoon or in the morning?

MARY: I would say about 6:30 at night.

LJ: And what was the purpose of coming over? Were you going to baby sit that night, possibly, or just visit?

MARY: No, I go there often.

LJ: For supper?

MARY: Yes.

LJ: Was Howard home at the time?

MARY: No, he wasn't.

LJ: He was away when you got there at approximately 6:30?

MARY: I think he was away; yes, he was. Rose had told me he was going up to the Pocono Mountain area.

LJ: She told you that in the afternoon?

MARY: When I arrived there. He told her that when he left.

LJ: But he had already left. Fine. Now this was about six o'clock at night. Was Rose in the house with you?

MARY: Yes. Rose was resting. She was over-tired, and she also heard the knock on the door.

LJ: And she did not get up?

MARY: She didn't answer it. I did.

LJ: Were the children asleep?

MARY: No, the children were awake.

LJ: Do you remember where they were at the time?

MARY: My nephew, Ricky, he was there; he heard the knocking. This boy is nine years old.

LJ: And would you say, when you answered the door—first of all, do the Menger children, when they hear somebody knocking at the door, usually go to the door, too, as a lot of children do?

MARY: Yes, they do.

LJ: Did they this time?

MARY: No, they didn't. Ricky was with me when I answered the phone. He also talked to his father on the phone.

LJ: Now, just let me get this straight a moment. This phone call came prior to the time that the pipe arrived with Howard? Or after?

MARY: After. About 10 minutes later. It was 8:20 when he appeared at the door and 8:30 when I received the telephone call.

LJ: Supposing we forget the phone for just a moment, because we are trying to find out the time that Howard appeared at the door. Do you want to say something, Howard?

HM: (Laughing): I know what you are thinking, Long John, and it doesn't take ESP to see it either. I was just as amazed when I called High Bridge and asked Mary. This was just like a hunch. An experiment on my part. And I was truly amazed and I am still amazed, and I don't know what happened. I honestly don't.

LJ: If you will just stick with me and let me present some questions. . . .

HM: The reason for the pipe was to present evidence that I had been there and I looked around the station wagon, 70 or 80 miles away, and could not find this pipe. Then I started to wonder, "maybe I'd better call. . . . Did it actually happen?"

LJ: All right, let me continue, Howard, and we'll get back to you about it. Mary, when you arrived at your sister's home that night about 6:30 and she told you that Howard had gone to Pennsylvania—some place in the Poconos—did she tell you that Howard was attempting a contact that night?

MARY: I don't believe. I don't remember that.

LJ: Did you have any idea that Howard possibly would try an experiment in teleportation?

MARY: No, I didn't, to tell you the truth. . . .

LJ: We would like that.

MARY: (Laughing): I don't know much about teleportation, but there was another incident that I know to be true.

LJ: You mean another incident other than this pipe?

MARY: Yes.

LJ: Well, let's take the pipe and then we will certainly go

on into the other one. This particular night that the children are playing in the. . . .

MARY: In the living room.

LJ: And you were sitting there with them?

MARY: That's right.

LJ: And it was about 8:30 at night.

MARY: It was 8:20.

LJ: It was 8:20. I must compliment you, but of course Howard has trained you now, because Howard knows that dates and times are very important. Do you remember what picture was on television at the time?

MARY: Gee, I don't remember.

LJ: It is not of any great importance anyway, and I don't blame you for forgetting—for that was rather a shocking experiment, to say the least.

MARY: It certainly was.

LJ: When you heard the knock at the door—does Howard have a particular way of knocking?

MARY: He never knocks.

LJ: He just walks in.

MARY: Certainly, he just walks in. It's his home.

LJ: I am not familiar with living in the country, but don't you lock the doors there? Are they usually open?

MARY: At night when you go to bed you usually lock them.

LJ: The reason I say that is that if you come to my apartment in the city, you would have to knock, even if you were a member of the family. I think you can understand that—living in the city? We just don't feel it is safe to leave the door unlocked. So Howard normally would not knock?

MARY: Well, you know, the way the knock was—it was very insistent, just like a pounding, a steady pounding—it was very different.

LJ: Steady pounding, for a considerable length of time?

MARY: Yes, just like a banging away.

LJ: What was the reason? If it was Howard there, whether in the astral or physical body, or whatever it was, why did he have to continue to pound? Was the living room fairly close to the door?

MARY: Well, I did not answer right away. I waited a couple of minutes. I thought Rose was going to answer it, and also it was at the back door.

LJ: The back door?

MARY: The back door.

LJ: Now, you open the door and Howard was standing

there. What was your experience at that time? I mean, how did you feel?

MARY: The way Howard looked, I was shocked.

LJ: When you say, "looked," you mean how he was looking at you—his eyes—his appearance?

MARY: He looked right through me. Right through me. It was very strange.

LJ: What was his next movement?

MARY: His hand just seemed to come out automatically and he handed me the pipe.

LJ: In other words, the pipe was in his hands—or did he take it out of his jacket?

MARY: No, it was in his hand.

LJ: Was the pipe warm?

MARY: No, it didn't feel warm.

LJ: In other words, Howard is smoking a pipe at this very moment. Will you feel that pipe for a moment (she feels the pipe)? That feels warm, doesn't it?

MARY: No, it didn't feel that way. It wasn't lighted. It was filled with fresh tobacco, but the pipe was not lighted.

LJ: And you said nothing?

MARY: I was shocked. I didn't say anything. I just accepted the pipe and inside the door I went.

LJ: What did you do at that time?

MARY: I called Rose and I said, "Gee, a very strange thing happened," and told her about it. Then 10 minutes later I got the phone call.

LJ: What was Rose's reaction to this strange phenomenon?

MARY: Well, I don't quite know. I guess she felt strange too. She didn't know what was going on, either.

LJ: Why did you feel so strange?

MARY: Uhmmmmm (thinking). . . .

LJ: In other words, let us assume this just for the moment: you knew that Howard was up in the Poconos?

MARY: Yes.

LJ: Well, isn't it quite possible that you, if you arrived at the home at 6:30 at night, or approximately that time, and you told us it takes one and a half hours to get back from the location that he was at if he was in the Poconos at the time—because you got the call 10 minutes later—isn't it possible that you would assume Howard had returned from this trip in the Poconos, and just stood at the door for a moment and handed you the pipe because, maybe, he had a little trouble with the car?

MARY: No, because he said he would be home in an hour and a half.

LJ: When did he say he would be home in an hour and a half?

MARY: When I got the phone call, ten minutes later.

LJ: Yes, but you haven't got the phone call. You are still at the door. You came from the living room . . . you were watching television and Howard is staring at you, looking right through you, and he hands you this cold pipe. Why did you feel this was anything mysterious?

MARY: If you had seen him, you would have thought it was mysterious, too.

LJ: That is a good answer. I was not there, and I didn't see him. Now let's get to the telephone call.

MARY: That was 10 minutes after I saw Howard at the door.

LJ: You picked up the phone?

MARY: Ricky, my nephew, did.

LJ: Did he talk to Howard?

MARY: Yes, and then I took the phone—I guess about a couple of seconds after he. . . .

LJ: Did Ricky say, "Hello, Daddy?"

MARY: He didn't know who it was. He said, "Who is this?" and he couldn't give him the message straight, because it was too far away. It sounded far away and he couldn't hear him good.

LJ: It was a poor connection?

MARY: Yes.

LJ: Then you took the phone, not Rose?

MARY: Yes, and it sounded very far away to me. It was a poor connection I would say.

LJ: Do you remember the conversation you had with Howard at that time?

MARY: He said, "I want to verify this: was I there?" And I said, "Yes, you were just here about 10 minutes ago." And he said, "I just wanted to verify it. I am up in the Poconos and will be home in an hour and a half."

LJ: Did Rose talk to him?

MARY: No, she didn't.

LJ: Didn't you tell Rose? By now you told her you had received the pipe?

MARY: Yes, and I showed her the pipe.

LJ: Wasn't she anxious to talk to Howard?

MARY: Yes, she was.

LJ: And you didn't give her the phone?

MARY: She was resting.

LJ: You mean she was able to rest after you had told her of receiving the pipe in this strange manner?

MARY: After he had teleported himself, I went into the bedroom and told Rose about it and then. . . .

LJ: The phone rang?

MARY: The phone rang and I went and answered it and then I went and told her again that Howard was on the phone. She was resting.

LJ: She was still resting after this?

MARY: Yes. Rose, however, did make a phone call that night. She called a friend up in the Pocono Mountains. His name is Joe Bozac, and she called him to ask if Howard was there. And Howard, you were not there, were you (Howard shakes his head)? Not in that area? He was not there.

LJ: Pardon me a moment, Mary; when did Rose decide to call Joe Bozac?

MARY: Right after I received the phone call and I had told her and she right away called Joe.

LJ: Now, Howard came back in approximately two hours, whatever the length of time may be, from the Poconos, and you possibly discussed it with Howard?

MARY: Yes, and he asked again, "Was I really there?" And I said, "Yes," and described what I saw and how he handed me the pipe; his expression and how he acted. I told him how very strange I felt.

LJ: Now, after Howard arrived and you sat in the . . . whether you did or didn't I can understand it, it isn't fresh in your memory . . . whether you had a cup of coffee or tea and talked . . . about how long after that did you go back to the Poconos? Was it 15 minutes or a half hour later when you and Rose and Howard went back to the Poconos?

MARY: At 10:20 Rose and I and Howard and two other persons left for the Poconos to go to Joe Bozac's house, or where he was staying.

LJ: What happened when you got there?

MARY: We just talked about it.

LJ: You mean you talked about the teleportation?

MARY: That's right.

LJ: Howard, when you arrived home, you were shown the pipe?

HM: Yes, and I couldn't believe it.

LJ: Was it the pipe you had with you?

HM: Yes, it was the very same pipe and it had unused,

unlighted tobacco in it. And the funny part of it was that it WAS cold.

LJ: I would just like to say to the people on the Party Line at this time that I have had the opportunity of knowing Howard about three months, and Howard is a big man in the pipe department. So, it is certainly not unusual for him to have two pipes in his station wagon—and if he told me he was carrying six, I would buy it. So the fact is, I know some people at this time are wondering, why don't I ask him if he would find a pipe in his station wagon as well as at his home. This could be quite possible with Howard because he could present five pipes through teleportation because he carries a lot with him.

HM: I was very excited about this, Long John, and I wanted to get up there and see Joe Bozac right away!

LJ: Why did you come back?

HM: To get these people and get right back again.

LJ: To get your wife, Mary, and a couple of other people? Why did you want them?

HM: Well, the more people I can get to understand this thing—though I do not understand it myself yet. . . .

LJ: What did Joe Bozac think of it?

HM: I don't know. I don't think it is up to me to say. I think it is up to Joe to say. I know Joe is interested in this type of thing and I thought I should give him as much information on it as I could.

ROSE: I made a phone call between 8:30 and 9:00 P.M. to Joe Bozac and at the same time on the line he received another call on another line from Howard; in other words, he picked up two phones and I was on one line and Howard was on the other.

LJ: What does that prove?

ROSE: That he, Howard, told him that he was in the area and wanted to see him.

LJ: And did Howard go to see Joe Bozac?

ROSE: I don't know. Did you, Howard?

HM: No, I came home first and then brought these people up with me.

LJ: Well, then that didn't prove anything, Rose—that Howard was in the Poconos at that particular time.

ROSE: I think it would, because Joe knew he was close by. Howard told him.

LJ: Because of the voice?

ROSE: He called from the diner.

The reason I transcribe this particular section of taped program is to present to the reader the type of interrogation that most of my witnesses were exposed to. Bear in mind that this was a friendly and congenial interrogation; outside of Long John Nebel, however, other interrogators and investigators were not so friendly. Most questioning was designed to "break" a story, rather than present a story.

With your indulgence I would like to present another case. This will be briefer, but I think you will understand why I am taking valuable space in presenting these reports:

January 10, 1957. A taped telephone conversation between a "Mr. X" and Long John. "Mr. X" is telling Long John about an experience he had with extraterrestrials. "Mr. X" was a physicist for 15 years, and is now a New York businessman:

Mr. X: So the five of us went out and he took us through very rough terrain. The underbrush was kind of high.

LJ: Five of you went?

Mr. X: Yes. The five of us went (Howard, Rose, "Mr. X," a young lady and her mother), and Rose Menger pointed out a glowing light between the trees.

LJ: Did you actually see this light?

Mr. X: Yes, definitely. It would get bright and brighter. It was a very slow pulsation. It would take about 15 seconds to grow dim, and another 30 seconds to get brighter again. It was pulsating at about that rate.

LJ: How far away was this light—from where you were standing with this group of people?

Mr. X: About 200 or 300 feet away. And it could be seen only through the trees. We were in an open clearing about 50 feet in diameter, and at the end of this clearing were the trees and it was through these trees that we saw the light.

LJ: What did you do after that?

Mr. X: Howard Menger suddenly said, "Wait here," and he walked off toward the light. He didn't go very far. It must have been about 40 feet, and then he stopped and we heard two male voices talking.

LJ: Was he in a clear area?

Mr. X: No, he walked right into the trees. He was probably about 15 or 20 feet into the trees.

LJ: Did you have any flashlight or anything?

Mr. X: Yes, we had a flashlight to get to the location.

LJ: Did you have the flashlight on as Mr. Menger walked toward this group of trees?

Mr. X: No, he asked us to turn it off.

LJ: Now you heard two voices, one that you recognized as Mr. Menger's—and do you feel the other voice could possibly have been Mr. Menger's?

Mr. X: No, it had a different quality to it. It was more sing-songy than his voice.

LJ: Do you have any recollection of the conversation?

Mr. X: Well, I couldn't make out any words. I listened as acutely as I could. I wanted to hear what was going on. I couldn't make out any words at all. And this conversation went on for at least half an hour.

LJ: Half an hour. And the four of you were waiting for Mr. Menger to return to the group?

Mr. X: That's right. While he was talking we were facing Mr. Menger and this person—whomever he was talking to.

LJ: You could not see him, could you?

Mr. X: You could see a silhouette, a dark shadow.

LJ: Do you know it was another form there?

Mr. X: It looked like two forms. I could not distinguish any facial characteristics. But I could pick out two forms.

LJ: Was the other individual taller or shorter than Mr. Menger?

Mr. X: I would say the other individual was slightly taller.

LJ: What happened at that point?

Mr. X: While we were looking in that direction we heard somebody walking along the edge of the trees, walking toward our right, and stopping about 90 degrees to our right. Then we heard another person walking. We could hear the underbrush crackling and they walked around almost to the rear of us among the trees. In other words, we were surrounded. It was kind of eerie at first, and I felt we were being observed. We couldn't see them, but I am sure they could see us because we were in the clearing.

LJ: You use the term, "them." Do you mean to imply there were more than one?

Mr. X: I am quite sure there were three of them, because you could hear two of them on our right, and then Howard Menger came back and he said, "Gee, I'm awfully sorry, I know how you feel, but I can't take you up there."

LJ: Did he give you a reason?

Mr. X: No, he didn't give a reason and he looked very, very disappointed. I am sure he wanted us to see what was going on.

LJ: Did he imply that these people were from outer space?

Mr. X: Yes. He did say there were three people: two men and a woman.

LJ: Did he explain to you what the conversation was about, what message they had?

Mr. X: I asked him what they said, but he said, "I can't tell you what they said; I would like to tell you but cannot," and it seemed to be something very personal.

LJ: What happened after that?

Mr. X: We just stood there and he (Howard) said, "If you get an impulse or telepathic message to go up there, go ahead." Then the young lady started walking and she walked up into the area, and I think she looked around there. She couldn't find anyone so she came back. And, as we were standing there, again we heard the two people on the right of us walking in the underbrush. I guess they were just walking back, and we just stood there for another half hour and nothing happened—just hoping against hope that we would get to see these people. And nothing happened.

LJ: Did you return home?

Mr. X: And then, Howard Menger said, "Just a minute," and he walked up again and again he spoke to somebody. Then he came back and said it was no use, that they were leaving. And after that we left the place.

LJ: That was the end of the night?

Mr. X: Yes.

LJ: Could this have been "set up?"

Mr. X: Well, we had a difficult job finding that spot. Howard Menger had not been there before. It was obvious we were lost when we started. And we had a lot of difficulty getting to this spot. We were going through underbrush and actually had to fight our way through. I am quite sure he had never been there before.

PART 5

DIET

One of the most outstanding characteristics of the people I have met from other planets is their good health. Their fine clear skin, bright eyes, and alert mannerisms are the result of abounding good health. They do not suffer from our many diseases, our colds, headaches, constipation and other common but distressing maladies. Therefore, they do not need doctors to cure diseases or alleviate pain; their approach to health is a way of life which stresses prevention rather than curative medicine. One of the main physical reasons for their health and longevity is their eating habits.

Diet not only includes the food that is eaten, but how the food is processed and prepared, and the type of soil from which it derives its nutrients.

The atmosphere on other planets, such as Venus, for instance, is not contaminated by industrial smog, atomic radiation, fumes, gases, offensive emanations from garbage dumps. The soil is healthy, and dust particles precipitating to the soil are equally free from contamination. On Venus there are magnificent, healthy fields of wheat, the kernels of which are three times the size of ours. These fields are tended by those who love to serve in that capacity. Venusians grow all sorts of grain, fruits and vegetables, including cabbages five times the size of our terrestrial version. After they have used a field for a season or two, they rotate the crops to another field, leaving the old one to rest and regenerate naturally. They do not scarify or denude the land of all vegetation and then leave the soil bare to the elements until the next planting. They permit unused vegetation, such as corn stalks, to remain on the field to decay, forming new food for the soil.

No fertilizers or sprays are used—they are not necessary when soil is healthy. We foolishly rake our lawns of all leaves in the fall, when it would serve the grass better if the leaves were left on the lawn to nourish the soil naturally. A good example of nature at work is a forest floor, where rich, dark organic humus enriches the soil each season for the tiny seed that finds healthy nourishment in its fertile womb. The large farms, where much of the food is grown on a huge scale for the people of various planets, are housed in long buildings, some miles in length, in which the crops are grown with scientific control in jelly-like substance containing the natural balance of vitamins, minerals and other necessary elements.

The buildings are constructed of a natural stone base with transparent tops. In the open fields and in the farm-factory buildings all work is done with advanced types of equipment, quickly and efficiently.

Botanists, chemists and food technicians supervise the operation, work out formulas, at the same time continuing their research. Food grown in this manner is not subject to parasitic or insect attack, although, from what I could gather, there is no pestilence on Venus.

After the food is harvested it is taken to another building where it is put into huge vats of liquid and allowed to seep for a few minutes. It is then lifted out of the vats and put onto conveyors. The moment it is transferred to the conveyors a bluish-white ray is beamed on the food for a second; then it continues on the conveyors to waiting vehicles where it is loaded. The processed food is then transferred to another building where it is stored, ready for use or shipment.

The fresh fruits and vegetables are immediately consumed by the people. They go to a commissary, select the food they need, then take it home. At home they can process it or cook it in their own kitchen. They have no refrigeration, since there is no need for it; moreover they do not believe in shocking the digestive system with extremes of hot or cold in food or beverages.

The processed grains will keep indefinitely. They do not bleach, spray and degenerate their whole grains. Their bread is really "the staff of life." It is dark, moist and rich in whole grain goodness.

My friends, our diseases on Earth are a direct indictment of our agriculture and food industries.

Their beverages consist of fruit and vegetable juices and warm drinks made from roasted grain. They serve vegetable soups of all kinds. Their sweets are natural fruit sugars and nutmeats. They make fruit jams of natural sweets, such as honey or natural syrups, with no preservatives added. The spreads are usually made from nutmeats or vegetable oils.

They do not boil, fry or overcook foods. The food is cooked instantly from the inside out without destroying the vitamins.

It is a shame that in our civilization we find it necessary to spray our foods, bleach them, kill them, render them void of minerals and vitamins, and then run to the corner drug store for our quota of pills containing some of the lost food elements—however the synthetic vitamins are not the same as those found in proper balance in foods grown in healthy soil.

Milk is given neither to adults nor babies.* Infants are breast-fed, and after weaning are given bland vegetables and fruit in puree form. Venusians do not raise animals for milk or meat. Animals are left to the natural balance of nature and there is no over-production of livestock. They roam free. It is we on Earth who have over-sexed, over-sired, over-studded our animals to ridiculous surpluses.

Our human system is a living example of bio-chemical warfare, due to drugs, injections, devitalized foods, tobacco and alcohol. These are the poisons of our world: over-consumption of animal proteins, candies, cakes, pastries, white flour, white sugar, alcohol, tobacco, drugs, carbonated drinks, ice cream, and all foods treated by poison sprays, poison gases, commercial fertilizers, bleaches, etc. Those rugged individuals who do attempt to eat natural foods must pay three and four times the normal price in our so-called health-food stores. Thus the average family cannot afford to buy the proper foods unless they can grow it themselves.

It would require a complete change-over in our agriculture and food-processing industries to give the people natural foods in unadulterated form. In the long view it would save a lot of money, but big money interests in established food processing, slaughter house, commercial sprays and insecticide industries, the pharmaceutical houses, the tobacco companies, would suffer huge losses during reconversion. Such a reconversion would come about only by enlightening the people and putting some teeth into The Pure Food and Drug Act, which continues to be ineffective against the overwhelming testimony of the cumulative effects of the addition of retarditives, chemicals and poisons to our food by manufacturers and processors every day.

Which brings us to the subject of chlorination and fluoridation of our drinking water. Natural fluorine does prevent tooth decay when taken naturally in foods; but chemical fluoridation of our water supply is a crime against

* The calcium content in milk is not always readily assimilated by the body, due to the imbalance of phosphorus which precipitates the calcium in the digestive system, rather than making it available for bodily use. Cow's milk is fine for calves, but not for babies. If it must be given to babies, orange or lemon juice or acidofilac should be added.

mankind. Are you aware of the fact that the addition of sodium fluoride to our water supply can affect our powers of reasoning by drugging a certain area of the brain which makes an individual more submissive and docile, much like the surgical effect of prefrontal lobotomy? This latter atrocity has been perpetrated upon victims of the Nazi regime and more recently upon unsuspecting patients in mental institutions.

This type of chemical and surgical warfare renders the populace submissive and easy to control. Sodium fluoride in your drinking water will, after a year's use, affect the rear occiput of the left lobe of the brain, where there is a small area of brain tissue responsible for an individual's power to resist that which is alien to his basic good. If these words are shocking or strong, they are written only in the hope that they will awaken people to the vast scope of the conspiracy (the accumulative mass of negative thinking and acting) which assiduously feeds into the public brain via radio and TV a negative approach to life and thus undermines the dignity of the individual and dulls his spiritual perception.

Can you imagine what would happen to the people of New York if all of them consumed water tainted with sodium fluoride for a period of years? Surgery and drugs are extreme emergency treatments and should be employed sparingly.

Dr. Carlton Fredericks of radio station WOR mentioned in one of his programs that "The physician of tomorrow will be the dietician, and conversely, the dietician of tomorrow will be the physician.

The physician of tomorrow will attack the problem of poor health on the proper battlefield: the blood. If the blood is made healthy through proper and therapeutic diets, it can withstand attacks. I was given information about proper diet for my son, who was, at that time, suffering from brain tumor and cancer.* It was also recommended that I see a Dr. Thomas (a true physician of tomorrow) who had cured more than 20 cases of cancer by the application of proper diet and the administration of optimum amounts of minerals. I went to this fine gentleman and asked him to help my son, whereupon he informed me he had given up the practice of medicine in favor of bio-chemical research in soil nutrition.

* It was suggested that I feed my son fruit and vegetable juices, minerals, fresh fruit and vegetables, such as apple-banana combination, cabbage-parsley-carrot combination (raw). The idea was to keep the blood alkaline at all times. An acid blood is a perfect incubator for disease germs.

He said he was tired of being criticized by a powerful medical organization because he did not overcome cancer by orthodox methods—although doctors had come to him to study with him and learn his techniques.

One time, when his products were at issue with the court, the judge was presented with evidence, including a case history of a patient who suffered from bladder cancer. The patient, who was present in court, should have died two years previous according to hospital records, and this fact caused the judge to remark, "To stop his (Dr. Thomas') work would set this science back one century. Each century produces a genius, such as Pasteur, Koch, Einstein; and Dr. Thomas is no lesser light." The judge then dismissed the action.

Dr. Thomas had read to the judge for one hour in his chambers the histories of a great number of cases treated by doctors and hospitals in the treatment of which Dr. Thomas' products and methods were used. He explained that during his biological research between 1908 and 1915 he had discovered a peculiar microbe which was constantly present in cancerous tissues. When the blood, however, had been regenerated, the microbe had ceased to grow. He explained how he had subsequently commenced his research into the production of a complete organic fertilizer which would produce better balanced foods from soil.*

Dr. G.H. Earp Thomas (now in his 80's) agreed to help my son if I could obtain the services of an A.M.A. accredited doctor to cooperate with him and accept his techniques. We asked a doctor in the vicinity to cooperate with Dr. Thomas, but he, like others, refused, even after seeing documents and affadavits attesting to the elderly doctor's work and success with cancer patients.

My son was taken to Dr. Hoxsey's Clinic in Pennsylvania for examination and possible treatment. The internal organs were found to be too far gone (especially the liver) which made treatment inadvisable at that point. I believe that if my son had been taken in time to the Clinic, a cure could have been effected—or at least an arrestation. And I am positive that if Dr. Thomas had been allowed to treat my son, the cancerous condition could have been halted.

* In 1953 researchers at Columbia University discovered that our soil was deficient in amino acids and the trace elements, which, if restored to the soil, would eliminate disease in plants, animals and man.

So not being allowed to practice his techniques in the cure of cancer, Dr. Thomas decided he would attack diseases at their source: in soil nutrition and in the food we eat.

Through his research he developed a type of organic fertilizer under the trade name of "Organo" which replenishes the soil with minerals, vitamins and the trace elements. He also invented the "Digester" which digests garbage and makes a perfect organic fertilizer from it. I also learned, quite recently, that Dr. Thomas has been experimenting for many years with growing cultures and small plants in jelly. The jelly is completely safe from bacterial attack and the plants are healthy and vigorous. He recently showed me such a plant in a test tube growing in the gelatine substance. This same, or a similar process, is used on other planets. Dr. Thomas is a true scientist and researcher, a hundred years ahead of his time, and I hope that some day soon his genius will be recognized.

PART 6

A NEW CONCEPT OF NUTRITION

By G. H. Earp-Thomas, Earp Laboratories, High Bridge, N.J.

Author's note: I present the following because I consider it a "must" for those interested in their own good health, their children's health, and the eventual good health and nutrition of all mankind. This is an excerpt from *Volume Two* of G. H. Earp-Thomas' publication titled *Cause of Disease Is Overcome By a New Concept of Complete Nutrition.*

Dr. Thomas, graduate of Duendin University, and Trinity University, inventor, lecturer, author, and biologist of international renown, developed Farmogerm, the first successful pure nitrogen culture for the inoculation of legumes. He developed the first system of composting with bacteria in compost cells—a kinetic soil analysis system. Among his other accomplishments have been: patenting the first composter for disposing of household refuse; the first complete organic fertilizer; the first preparation of trace minerals for deficiency diseases; the first pure composite culture of legume bacteria; the first lactic acid culture to improve ensilage in silos; a new complete food for chinchillas; the first complete organic fertilizer made from fish and from vegetable wastes. All of these developments and inventions have been used successfully both domestically and abroad, some of them for many years.

He also made a new food to supplement meals and was the first to include all 32 factors needed for nutrition. He also introduced to the medical profession a new concept of disease, and prepared the foods including the trace elements to found a new system of medicine, Bio-Chemical Therapy, and was granted a diploma for so doing by a medical college. He is also a noted radio and television speaker—Howard Menger.

More than 6,000 hospitals in the United States operate 24 hours a day. These hospitals have a combined capacity of approximately 9,000,000 patients. Yet ailments such as heart afflictions, cancer and nervous disorders are on the increase. Approximately 165,000 licensed medical men and probably four times that many unauthorized healers daily try to control disease. But the avalanche of death continues to march on.

This nation has laboratories by the hundreds, experimental stations, clinics and countless health departments functioning as efficiently as an endless stream of dollars can make them. Great foundations for human betterment, with branches throughout the civilized world, are constantly issuing glowing reports of their work. Thus the public is led to believe that the health millenium is just around the corner. But always—always something interferes with the full fruition of these dreams. The death toll continues to mount. The combined efforts of all these agencies appear to be of no avail. The battle for health is a losing battle. Each year finds the Grim Reaper more and more triumphant.

Such facts are not mere suppositions. Neither are they an unexpurged alarmist's propaganda. Figures, checked and rechecked, tell the gruesome tale. No amount of fine words—no amount of Polyanna speeches can conceal the dismal truth. So far as civilized health is concerned science has failed—failed miserably.

There is a reason for this failure. Providence cannot be blamed—Providence has not cast an unfriendly eye upon this land of ours. Not at all. The failure is the direct result of science taking the wrong road approximately a century ago. Theories and rules were made then without sufficient laboratory research. And many of these theories and rules made then since have been proven unsound and unreliable. Yet science has gone a long way on this false road. It resentfully refuses to admit its mistakes and to retrace its steps. It will not wipe the slate clean and begin again. Instead, science tries to harmonize its latest discoveries with the incongruities of its false position. And such harmonization cannot be successfully accomplished.

Early Physicians Valued Minerals

About one hundred years ago the physicians were beginning to understand that the human body needed various kinds of minerals in order to maintain health. But it was mere speculation as to what those needed minerals were. They believed the body needed lime for the bones and teeth—or technically, calcium. They believed that the nerves needed the mineral phosphorus. But just how the body utilized these minerals they did not know. In those days—

calcium was ordinary lime. Phosphorous was ordinary phosphorus. What these physicians did not know, what they did not realize, was that the minerals, the calcium and phosphorus and others must undergo a definite change in character before they can be utilized by the human body. They did not understand that minerals, in the ground, are vastly different from those minerals of the same name the human body receives in the food that is eaten. Ordinary calcium may be splendid to whitewash a cellar or plaster a wall—but it must be vastly changed before our body can use it for bones and teeth. And the human body alone cannot make this change. Yet even today science clings to that outmoded belief that in some instances the body can use such dead minerals.

Iron Not All Assimilable

Of course it has been demonstrated that if dead iron is injected into the blood of a human being a definite reaction is produced. The blood becomes flustered from the presence of this outside element and science erroneously draws the conclusion that this mere excitement of the blood is absolute proof that such a dead iron is utilized by the body. Unfortunately a microscopic examination of human blood fully ten years after such injection will show traces of unassimilated iron. The rest is eliminated from the blood. However, it does aid organic iron to enter the circulation to make haemoglobin. It is known that inorganic iron does pass into the blood circulation and its only value is that it aids in the passage of any organic iron that may be present in food. The organic iron joins to form haemoglobin of the blood, but the inorganic iron does not and is eliminated from the blood as useless in the wastes of the body. What is more, no laboratory experiment today has shown that the iron of a nail, taken internally, can be changed by our body into the iron that our blood can use for nourishment. In theory such a transformation in our body takes place. In actual practice, positive proof is lacking.

Physicians, a hundred years ago, did not understand that chemical changes in the minerals were necessary. To them calcium was calcium—whether it came from a lime kiln or from the food we ate. Thus science started off on that false road which leads down the passages of time into today.

In spite of the increasing knowledge regarding nutrition

orthodox science steadfastly refuses to acknowledge that its original conception of minerals is wrong.

Many nutritionists, however, soon realized this error. But their suggestions for making a distinction between these two types of minerals for the most part fell on unhearing ears.

Organic Mineral Compounds Superior To Inorganic Minerals

However, these same nutritionists continued their researches and these two kinds of minerals were classified. These minerals—as they come from the ground, such as iron, calcium, phosphorus, silicon, sodium, etc.—were designated "dead" substances. This means they are not suitable for food. They are now called officially "inorganic minerals."

Those same minerals, after they have made the trip through the digestive processes of a plant or a microbe, and have been chemically changed so that the human blood may absorb them, are called "live" substances. These minerals are suitable for food. They are now called officially "organic minerals."

But even so science acknowledges no such difference. In a late issue of "A Guide to National Advertising" issued by the National Better Business Bureau is the definite statement:

> "Some advertisers have represented that organic mineral compounds are better than inorganic and that organic minerals are the 'necessary' or 'proper' form in which minerals must be obtained in the diet. In general, there is no established scientific basis for this premise and in some cases, the facts are quite the reverse. The availability of a mineral for human nutrition does not depend upon whether it comes in an organic or inorganic state but upon solubility and other factors still under investigation."

In light of the above it is well to bear in mind the testimony of the great Dr. Charles Mayo of Rochester, Minnesota, offered to a Congressional Committee in the City of Washington, D.C., not so many years ago. He gave it as his opinion that there was absolutely no relationship between proper diet and disease.

How that idea has changed in the light of experience and

laboratory tests during the past decade! Note: I made these notes many years ago.

Now it is recognized that proper diet may not only relieve practically every type of disease, but in most cases the correct food may absolutely prevent the human body from "catching" ailments. I will enlarge on this in later pages from recent research.

Even today no sensible scientist would tell you to place an iron nail in water, let it rust and then drink the rusty water in order to obtain iron for your blood. They would know that such iron would merely pass through your body without being absorbed in any manner. What is more, science knows that iron extracted from plants—after it has been chemically changed from "dead" iron—will not melt. It cannot again be formed into a nail. No process known to man will render it "dead" again.

Not only that, but no sensible scientist would tell you to break off a piece of plaster from the wall and swallow it in order to get lime for your blood, your bones and your teeth. Science knows that such lime—such "dead" lime—would be absolutely useless in the body.

Yet science will not admit officially that as far as the needs of the human body are concerned there is any difference between organic and inorganic minerals.

It is certainly time the public was made to understand this mineral situation. It is certainly time for the man in the street to be in a position to judge just what type of minerals his body needs. It is certainly time for everyone to be informed that all minerals to be used by the body must first be digested by fruit, vegetables, grain or other vegetation. Reliance upon inorganic minerals, either in medicine or food, is not only unwise but it is also dangerous to health. No wonder that science has made a botch of its battle with disease. It is standing on a false platform. No wonder there is not one single disease in the tens of thousands recorded in the medical books after which can be placed the words, "perfect cure" in the same manner that Q.E.D. may be appended to a successfully worked out theorem in geometry.

Common Cold May Be Just A Symptom

The National Better Business Bureau, ever alert in its efforts to prevent the misuse of any wording respecting the advertising of a remedy, concedes in its printed booklets that not one single remedy known to man can be used unqualifiedly to denote a positive cure for a single, solitary disease.

Even the common cold which incapacitates more people than any other ailment in the United States is still shrouded in mystery.

The cause of a cold is unknown. Yet it is said that one out of every two persons has at least two or three colds a year. How the cold enters the human body is still wrapped in mystery. What the cold does, after it arrives inside of us, is only the vaguest sort of speculation.

And yet, the cold is strictly a product of our civilization. The savage in the tropics and the Eskimo in the frigid zone do not know the miseries of the cold. Only when they become "civilized" do they experience these attacks.

Not only is the common cold an unsolved mystery but there is no known remedy, so far as science is concerned, for such everyday afflictions as arthritis, rheumatism, heart trouble, kidney disorders and liver ailments.

Even the word "constipation" has come under the ban and cannot be used over any reputable radio broadcasting station. Many physicians declare there is no such thing as constipation. Yet how otherwise adequately to describe the clogging up of the human digestive system with fermented waste material is something that would take the collaboration of an Einstein and the proverbial Philadelphia lawyer. Some authorities claim—and there is no reason to doubt them—that practically every civilized adult human being suffers in some degree from constipation.

It is time someone explained just why these unaccountable upsets occur in the human body. It is time someone forgot symptoms—forgot nervous disorders—forgot arthritis, diabetes and the multiplicity of troubles that affect our vital organs. It is time someone explained in simple language the exact underlying cause of all ill health.

Vivisection Has Availed Little

For years we have listened to the controversy over the morality of experimenting on live animals in the laboratories. Many conscientious persons claim that no good actually comes from such modern torture chambers. And considering results, these objectors to vivisection are not far from the truth. Science with all its laboratories cannot take to itself the healing of a single ailment. Science can help correct unnatural conditions in the human body but Nature alone can heal. Nature is the only healer.

Through countless ages, before science was ever old enough for its swaddling clothes, Nature alone took care of human beings and brought them successfully down through the centuries.

The elephant survives in spite of the fact that no Jumbo or any other distinguished pachyderm ever qualified as a healer. The horse is still with us and yet no horses have developed an intelligence that would have permitted them to take a degree in a scientific institution. This survival was not mere luck. This survival is a definite part of the scheme of things. Animals left alone conform to Nature's laws. Man is the only animal who seeks to defy them. Let us see how man has fared because of his defiance.

Now sickness and ailments were not originally contemplated by Nature in the future she pictured for man. Nature gave our bodies everything necessary to keep us well. But when man defied Nature's laws he upset the scheme of things. He tried to live his life according to his own ideas, to suit his own fancy.

The Unnatural Life We Lead

Nature intended man to live permanently in the tropics. In the Torrid Zone she supplied him with nourishing food in abundance. The climate was suited to the hairless condition of his body.

However, man preferred to journey. He left the tropics and set out for colder climes. It was a foolish migration. Man was not equipped for such changed conditions. His body lacked protecting hair to keep him warm. Naked, he was unable to bear the rigors of the winters. So clothing became necessary.

With the donning of clothing Nature's scheme for man's health was immediately thrown out of balance. The clothing shut out the free circulation of the air about man's body and this ever-changing air was prevented from reaching the many pores in his skin. The poisons from these pores could not now be easily dissipated. And excess poisons, held in the body, meant ill health.

Man needed more warmth—so he sought out caves in which to live and adopt fire as his own.

Nature intended man to dwell in the open, to sleep from sun to sun. But when man found fire, he obtained light. Then came candles, oil lamps, gas and finally the electric bulb. None of these methods for making light equaled that of the sun. Man's eyes soon became deficient.

Later man ceased to live in caves. He built habitations of wood, stone and brick and to prevent the entrance of the chill he made glass. Thus he shut out not only the fresh air but also by the use of glass he strained out the most healthful ultra violet rays of the sun.

Later man made shoes for himself to incase his feet. Thus he encouraged more ailments by closing up the pores in the bottom of his feet. The large drains in the soles of his feet had provided additional outlets for body poisons. Man ignored the fact that Nature had intended these poisons to be released into the ground. He ignored the fact that these pores were the larger vents in his body. Through the use of shoes he kept these poisons against his feet and ankles. Then he invented heels for his shoes to raise his feet out of the hot sands. And in putting this unnatural wedge under his heel, he threw his entire body too far forward. Thus he put an unnecessary strain upon his delicately balanced vital organs. Yes, indeed, man brought disease, ailments and discomforts upon himself. He refused to work with Nature and tried to work against her.

But even so Nature did not abandon man to his own devices. Nature continued to fight valiantly to keep ailments and diseases from his body. She provided in his blood an army of defense, hoping that man would keep this army always in first class condition.

It should be borne in mind that the human body is nothing more than a chemical laboratory. It is filled with little live creatures, which carry on the processes we call life. If the human body would in some manner destroy these little creatures, life would instantly cease. Of course, not all these little creatures in our body are helpful ones. Many have the power to actually destroy our health.

A Balance Condition Needed For Well Being

The well person is the person who has more of these beneficial little creatures than harmful ones. The state of your health depends primarily upon the balance existing between the beneficial organisms and the harmful organisms in your body.

In our blood float different types of these living beneficial creatures. One of these types, which are in the vast majority, are called red corpuscles. These red corpuscles are born in the marrow of our bones and in our spleen. Yes, the marrow of our bones and our spleen are little more or less than manufacturing plants for red corpuscles.

If we give to our body everything it needs to nourish these little red organisms we thereby enable our body to bring forth billions upon billions of these red corpuscles.

These red corpuslces, each one alive, only live from thirty to ninety days. They must be constantly replaced. They have their childhood, their middle age and their time of death just as the human body has its limit of length of life. It is the duty of these red corpuscles to separate the gases of the air. They collect oxygen from the air. These red corpuscles in our blood travel along our veins to our lungs. In our lungs they seize the air that we breathe in and take out the oxygen. This oxygen is then delivered by them through our arteries to all parts of our body.

This oxygen is used by our cells to burn up and make energy out of the nourishment our cells have received from the food we eat.

Floating along with these red little organisms in our blood are also white little organisms. These white little organisms are called white corpuscles. Their duty is different from that of the red corpuscles. These white organisms constitute our defending army. They are the fighters for our bodily health. And what fighters they are! It is their duty to trap and destroy any harmful germs that may enter our body.

Most of you know that the air that surrounds our body—the air that we breathe in—is filled with all manner of destructive organisms which science calls germs. Some of these harmful germs, when they find a spot that is to their liking, develop and multiply until we have a condition in our body known as diptheria. Other harmful germs develop and multiply until they show their presence in such number that science calls the condition typhoid fever.

That is the story of each and every bodily ailment.

The white corpuscles in our blood are constantly on the lookout for such invaders. And our contaminated air brings into our body these harmful germs in countless numbers every time we breathe in. When they arrive in our lungs—

and enter our blood, they are instantly set upon by the white corpuscles. And these white corpuscles seek to tear these harmful germs to pieces by digestion.

White Cells Great Scavengers

Under the microscope these white corpuscles in our blood show a ferocity unmatched by any beast of prey. They set upon harmful germs and tear them apart with all the speed and alacrity of a bulldog shaking a rag doll. They hunt in packs like wolves, surrounding the victim and rending it from all sides.

Now as long as these white corpuscles in our blood are kept in the best of condition—so long as their number is up to standard—they are capable of fighting off almost any number of disease organisms that may enter our body. Just remember that! If these white organisms are well nourished—if this little army of defense is well equipped through the food we eat with everything they require for their own purposes, we need have little fear that disease or sickness will gain the upper hand.

Yet, if we do not give to our body what it needs to keep this defensive army well equipped and vigorous—if we allow our body to run down, to be undernourished, then these defense battalions of white corpuscles in our blood must fight a losing battle and ailments will follow.

Disease is not a cause—it is a result. Disease is Nature's way of telling us that something within our body has gone wrong—that the delicate balance of our life processes has been upset. No malignant germ can long survive in a healthy blood stream. The white corpuscles would destroy it.

Food Now Incomplete

So long as the human body receives everything it needs in the food that is eaten, it retains fairly good health. Despite all the handicaps man, through his defiance of Nature's laws, has placed before her, Nature continues to fight for our health. So long as man ate good food, food that was filled with everything his body required for its life purposes, Nature continued to battle successfully. But man, during the past hundred years especially has been tampering with his food. He believed he could improve upon Nature's handiwork.

Man took the wheat as it came from the fields and to make it look better he stripped it of its skin which contained many of its most nourishing elements. He extracted and threw away the life element of the wheat—the germ life element which permitted the wheat grain to sprout. He found that the presence of this life element prevented the flour from keeping for years at a time. When he was through, the flour that resulted was a white "ghost" food with most of its nourishment discarded.

But man liked this new flour. He resented going back to the flour made from the whole grain of the wheat—the flour that man had eaten and thrived upon since the days of antiquity. History records that when a New Englander by the name of Sylvester Graham wished to give the people of his community the wholesome whole grain flour of ancient times, the good folk of Massachusetts rose in revolt. Our friend Graham was stoned in the streets—his mill was pulled apart and his home burned to the ground.

Man not only devitalized and demineralized his flour but he refused to eat the sugar in its natural form. He took the juice of the sugar cane, strained it through charred bone filters and removed the molasses. Yet with this molasses went most of the nourishing elements the original sugar contained. The result of this processing was the glistening white powder we have today—another "ghost" food that has helped mightily to drive good health from the human body.

Man peeled his vegetables and with the discarding of the peelings his food lost many of the nourishing elements Nature had stored in this skin and just below this skin to preserve it. Man also cooked his vegetables, throwing away the water in which they were cooked. Man never realized that many of the nourishing elements of vegetables dissolve in water and that often the most valuable part of his vegetables, from a nutrition standpoint, was allowed to run down the drain in his kitchen.

In a thousand ways man tampered with Nature's foods. He doctored them. He dried them. He covered them with preservatives. In fact the dismal story of man's abuse of his food would fill volumes. It is an epic in itself and will go down in history as an unwitting attempt of the human race to commit suicide.

Still, despite man's shortsightedness as to his own welfare,

Nature fought on. She provided him with ample fruits and vegetables that were rich in the essential elements the human body needed for proper nourishment. And as long as man could depend upon this unspoiled source of nourishment he was able, in a measure, to weather the storm of his own bungling handiwork.

Soil Now Incomplete

But the final straw that broke the camel's back came when man in his selfishness actually destroyed the fertility of the ground in which his food grew. That was the time that even Nature was unable to combat the catastrophe. The moment that man had so denuded the good earth that the plants that grew thereon—the fruit, vegetables and grain—no longer could obtain from the soil the nourishing elements that man needed for food—Nature seemed to throw up her hands and allow human beings to become overrun with ailments and disease.

That is the condition the civilized world faces today—the soil no longer brings forth the bountiful nourishment that Nature intended human beings to receive.

This is not the vaporing of swivel-chair specialists—the men who tell us that this earth will freeze solid in a few million years. This is not the prediction of crackpot investigators seeking for a momentary basking in the limelight of publicity. It is the sober conclusion of nutritionists who for years have watched with grave concern the despoiling of our lands. The truth of this is reflected in the reports of health authorities and social workers everywhere. So do not regard this statement as a hare-brained theory about what may happen some time in the future. It is a condition that is upon us—a condition that must be faced at once—today.

Food Loses Vital Factors

Authorities tell us that there is now a Nutrition Front in the United States. They tell us that in addition to our other troubles we must be prepared to fight improper nutrition. They warn us in unmistakeable language that much of the food eaten by the American people is poor food. They show by facts and figures that as a result of the wide-spread ignorance of the laws of health our physical and mental conditions are lamentable.

These same investigators announce that fully 45,000,000 persons in this country are undernourished. Think of it! Here we are, in the richest country in the world, the country that boasts of its millions upon millions of farms, of its wheat and corn fields which stretch from horizon to horizon, and yet, it is now disclosed that fully one third of our people are actually undernourished.

Malnutrition Causes Rejection Of Men For Army

Now this condition of almost universal undernourishment has been known to many investigators and nutritionists for years. Yet no amount of shouting on their part, no amount of "viewing with alarm" could rouse the powers that be to a full appreciation of the dangers of such a situation. It took an outside emergency to bring this wide-spread condition of undernourishment to the foreground. It took the discarding of tens of thousands of draftees to wake this country from its lethargy.

For example, during the months of June, July and August 1940, approximately 6,743 men applied for voluntary enlistment in the United States armed forces in the metropolitan New York district. These men sought to anticipate the draft. They felt that they were in such fine condition that the draft boards would be sure to select them later. Their main reason for volunteering was that they wished to pick the particular type of service which appealed to them. These young men, investigation showed, came from practically every community in this country.

Yet of the 6,743 who applied for admission into the ranks of Uncle Sam more than 2,195 were turned down. More than 25% of the rejected men had defective teeth. Some 21% had defective feet and 10% were deficient in hearing. Lack of weight cut off another 15% so that the total of rejections was well over 69% on account of so-called minor defects. The remaining 29% of the rejections suffered from various organic disorders.

And what happened in New York City is but a sample of what occurred in every recruiting district in the United States.

Not only that, but in the operation of the draft itself the percentage of rejections was even greater. And yet the physical requirements promulgated by the draft officials were in no sense strict.

Now, orthodox science could have blamed these unfortunate physical imperfections upon many things. They could have passed the buck. The bad teeth could have been ascribed to the lack of the tooth brush and tooth powder. Ill-fitting shoes could have been blamed for the flat feet. The cause of improper hearing could have been put at the door of the bang, clang and rattle of our modern life. Even the organic ailments could have been made to appear to be a direct consequence of a lack of proper medical care. But science did not pass the buck. Science in unmistakable language accredited these conditions to the almost universal undernourishment of the people of this country.

Are We Starving At A Full Dining Table

But do not get the wrong idea about this undernourishment. There is no starvation in the general sense of the term in the United States. These undernourished persons are not hungry. There is no lack of enough food. The real lack—the only lack—is enough of the *right kind of food.*

As one prominent official stated:

> "We in America are living in a land of abundance. There is more food than can possibly be used by the inhabitants of this country. Surely there should be no malnutrition in this land of plenty! But there is. Not only does malnutrition occur here but is so wide-spread that it has come to be recognized as a major health problem. Though it seems paradoxical, it is nevertheless true that a person might eat a sufficient quantity of food and yet suffer from inadequate nutrition. Quantity alone, therefore, is not the solution of the problem."

Again, in the same vein, one of the leaders of the United States Public Health Service stated in an article published in a recent issue of the journal of a prominent medical association: Note: This was written before 1942.

> "Without attempting to put any specific figure on the number of people affected, such data indicate that in all probability the nutritional diseases constitute our greatest medical and public health problem, not from

the point of view of deaths, but from the point of view of disability and economic loss, a fact about which we have been misled by the very low death rate and inadequate diagnosis."

Let us dwell upon these last words:

"about which we have been misled by the very low death rate and inadequate diagnosis."

Now who is responsible for misleading the public? Who is responsible for the "inadequate diagnosis?" Surely, science has cause to hang its head in shame.

That is why the time is ripe—as the oyster in Alice in Wonderland might say—to talk frankly about these matters. The time is most propitious to point out that there are understandable reasons why the majority of the people of this country are undernourished.

Calories As Heat Energy Not Enough

For years science has been drilling into the public that each one must receive, in the food that is eaten, a certain amount of energy each day. This energy was calculated by science in the terms of "calories." Very elaborate charts were prepared to show just how many of these calories—or heat units—were needed by individuals in various pursuits of life. Oh, the scientists spent plenty of time and much research on this branch of nutrition. They watched their test tubes, estimated the energy required to move our various muscles and finally emerged from their studies with their chart. Every type of employment has been analyzed. The figures show just how many calories are consumed when you iron, wash, sleep, chop wood or play mumblety-peg. In fact, these scientists could tell you to a dot how much energy the body uses to sneeze or gasp.

Now if the truth were known all these calculations were not worth the paper on which their summary was finally written. A calory is the amount of heat which results when a definite amount of material is burned. And when you read that your body needs approximately 2,500 calories a day it means that the food you eat should give to your body that much heat. But that it is not its greatest menace. The danger in relying on such a set of figures is that a person eating food containing the prescribed number of calories for a given occupation might be actually starving for the lack of enough proper nourishment. Food that contains large amounts of calories may give energy to the body but may fail to give enough real nourishment to the blood, the heart, the teeth and the bones. The calory system of nourishment is obsolete—is misleading and in these days of denatured, demineralized, devitalized, processed foods is positively an ominous peril. It should be discarded or amended.

In the place of the calory charts, a new standard of food values should come into being. Nutritionists have realized this for years. A new scale of nourishment values must be made not based on the energy a food is supposed to produce in the body but rather on the amount of real nourishment that a food contains. And the real nourishing value of a food must be calculated not only on the heat it gives but on its mineral and vitamin content.

Minerals And Vitamins Most Essential

Most folks know that the body needs approximately eighteen different kinds of minerals and some dozen or so different vitamins to properly carry on its work. Now science will agree that it is almost impossible for a person to eat any kind of food that will give his body enough of these minerals and vitamins and not get enough calories for all the energy that person may require. On the other hand you can eat food containing thousands of calories and not get a single mineral or vitamin, such as sugar.

In other words, calories are merely incidental matters. The proper solution of the nutrition problem is to see that your body receives each day everything your body needs for its growth and for the repairing of the parts your activities have worn. The true nutritive value of food must be based on the minerals and vitamins that food contains—not only on the speculative energy it may produce in your body.

It is said that one pound of fat will give to the body approximately 4,000 calories of energy. And yet—with all this quantity of stored energy, not a mineral—not enough vitamin—is included in that fat. If you would eat fat exclusively, you would die in a very short while of acidosis. Your body must have regularly certain minerals and vitamins in order to keep alive.

It should be remembered that each one of these eighteen or so different minerals has its important part in the health of a human being. If one single mineral out of the eighteen is absent—or if any mineral is present but in too small a quantity—good health is absolutely impossible.

Some of these minerals are used in large amounts by the body. The mineral calcium, for example, makes up the major part of the bones and the teeth. Likewise the mineral phosphorus is used in large amounts—the nerves and the brain are mainly constructed of phosphorus. The mineral iron is used very sparingly and yet it must be present if a person is to have good, rich, red blood. If it is not present in large enough quantities we are troubled by an ailment known as anemia. If not enough of the iodine is given to your body in the food that is eaten, a large gland at the top of your chest, just in front of your throat, starts to enlarge and your trouble is designated goitre. Government and American Medical Association figures show that some 30 million people in this country suffer from iodine deficiency.

Every mineral and every vitamin has its proper place in the scheme of things. And not one of these minerals or vitamins should be absent. And yet some of these minerals are found in the body in such minute amounts that they could easily be placed on the head of a pin. Even so, science for years did not appreciate the worth of minerals in the body as they relate to health. For example, science looked upon the presence of many of these minerals in our body as either accidental or unimportant. Science could not see that the mineral fluorine had any particular function. Now we know that if a sufficient amount of fluorine is not present, the calcium, silicon and other minerals in the teeth will become soft. Yet few text books mention fluorine at all.

The problem of strong teeth is a national one. And whether the universal tooth decay in this country is due to a lack in the food customarily eaten of calcium, phosphorus fluorine, silicon or some other mineral science has not yet determined. Like the average committee in a social club "reports progress"—science says that the cause of tooth decay is "still undergoing investigation."

Yet explorers and travelers tell us that savage tribes, living entirely on natural foods, have little or no tooth decay. Tooth decay therefore seems to be caused by the lack of something in the civilized human body that it should have received from the food that is eaten, including fluorine.

Malnutrition Cause of Tooth Decay

Not long ago the students of the University of California were examined by a competent board of scientists. This board reported that 95 students out of every 100 had bad teeth—or a total of 95 percent of something like 5,000 students. It should be remembered that these students at the University of California came from the better class of homes. Now the average for bad teeth in America is approximately 97 out of every 100—so the students of University of California had teeth that were just 2 percent better than the terrible average for the entire country.

At the same time this announcement was made from California a most enlightening story appeared in the newspapers. It was the report of two travelers who had just returned from the upper reaches of the Amazon River in South America. These travelers had examined hundreds of the natives of the jungle tribes and had failed to find a single cavity in their teeth. In other words, the chewing equipment of uncivilized natives in South America were perfect as far as upkeep was concerned while the teeth of civilized Americans are almost universally imperfect.

It is widely conceded by nutritionists that tooth decay very seldom, if ever, comes from the outside of the tooth. Brushing your teeth twice a day or three or four times a day will not keep them strong if your body lacks proper nourishment as far as minerals are concerned. And civilized food manifestly lacks these minerals.

Good Teeth Come From Good Food

Right in point is another late investigation that was conducted by a prominent Philadelphia scientist among the headhunters of South America. Two groups of these savages were studied by this scientist. One group consisting of 67 men had never had any contact with the outside world—with our so-called civilization. These men had never tasted civilized food and had depended entirely upon fruits and vegetables as they grow in the tropics. These 67 men—these savages—had perfect teeth as far as the condition and strength was concerned.

The second group of these headhunters—80 in number—had enjoyed the benefits of civilized food for a number of years. And here is the result of these two different types of eating. While the first group of 67 men who lived on natural foods had perfect teeth—the second group of 80—who ate civilized foods—had become dental cripples. Out of the total of 80 men of this civilized food group, 77—that is, all but 3—had bad teeth.

While comparing the bad teeth of civilized man with the good teeth of the denizens of the jungle demonstrates a phase of nutritional deficiencies in the United States—decayed teeth are really a minor manifestation of the ill health of the people of this country. In the medical books are thousands of ailments that might well have as their underlying cause the lack of this or that mineral. I only cite the cases of these savages to bring home to everyone that the cause of this country's universal tooth decay cannot be wiped out by tooth brush or tooth paste. Civilized food must be revitalized—it must have given back to it the nourishment that is missing. I shall later show how this may be brought about.

And this goes much deeper than making some improvements in our food-processing methods. If the food, as it comes from the field, is actually lacking the necessary minerals and vitamins no amount of improved processing will make that deficient food good food. Let us not forget that.

Food Now Lacks Vitamins And Minerals

Lack of vitamins has the same destructive results upon the human body as has the lack of minerals. Of course, vitamins are a new study. They are recent arrivals in the scientific field. More publicity has been given to them than minerals. Yet very few persons understand what vitamins do in the body. Very few persons know why they are essential.

A vitamin is not nourishment in the sense that minerals are nourishment. Vitamins do not help build the cells in the body—they do not create the bones, the tissues and the muscles as do the minerals. A vitamin resembles the spark in the engine of an automobile. Without the spark, the gasoline cannot be ignited to give power to the motor. And without the vitamin, the food we eat cannot be properly used by the body. In scientific language the action of a vitamin in the body is a catalytic one. Simplified it means a substance which permits two other elements to join together like hormones.

Vitamins are just as important as minerals in the field of

nutrition but they serve another purpose. Let me repeat, they are not nourishment. Their presence merely enables the body to take from the food we eat the nourishment it requires—provided of course the food has that nourishment. Many times civilized food is sadly deficient in that respect. And remember, if the food you eat does not contain enough protein and minerals to meet your body's requirements, all the vitamins in the wide, wide world will not give you good health.

The old expression that "pigs is pigs" does not apply to food—"food" is not necessarily "good food." Not only that but the character of the same food grown on the same ground changes from year to year. If the nourishment that each crop robs from the ground is not replaced, the food grown on that ground—while it may look exactly like the food that was previously grown—may not contain even one-half the nourishment of the original food.

Plants vary in minerals and vitamins to the same extent that the ground on which they grow varies in minerals. If there are many minerals in the soil the plants can use—the plants will contain plenty of minerals. If the soil is short on the proper minerals—the plants will grow, yes—but they will likewise be short as far as nourishment is concerned.

Poor Plants Make Poor People

Unfortunately, the whole country is eating this type of food that is short on nourishment. The soil has been so ruthlessly robbed that the plants—while they look the same—are all too many times delusions and snares as far as nourishment is concerned.

When the first Europeans landed on these shores of the Western Hemisphere—some 450 years ago—they found a continent that was far different from what it is today. The soil was fertile and self-sustaining. The ground was alive with a luxuriant vegetation.

Great forests covered the mountainous country and the heavy spring rainfall seeped into the soft ground and was held there for the hot days of summer. Only on the great western plains were there signs of soil weaknesses. Yet even in that semi-arid region, vegetation grew in profusion along the streams and rivers and wild Buffalo grass on the plains.

But man is a destructive animal. He thinks first of his own

temporary comfort. He gives little thought as to how his short-sightedness will affect the generations to come. So with his axe he denuded the hillsides and with his factories he polluted the waters of the streams.

In a short four and one-half centuries man destroyed the balance Nature had been building up in the soil for millions upon millions of years. Early settlers looked only at surface indications. They gave no thought to the intricate processes of plant feeding that took place down in the ground beyond their sight. Fertility to them meant crops—and they believed that land was inexhaustible in its ability to give and give without a return to that land of the things of which it had been robbed.

In this man made a big mistake.

Soil Fertility Depleted

The fertility of land must come from definite sources—from the minerals in that land and from the nitrogen, oxygen, hydrogen and carbon in the air and in the water. If the minerals are there—but not in a form the plants can use—no plant can possibly permanently survive. Man gave little thought to how this air and water was broken up into food a plant could use. He gave little thought to its importance. His analysis showed that plants needed certain elements to live and as his soil became less productive with each succeeding crop, he supplied chemical minerals to his land in the hope that his seeds would find in these minerals alone the sustenance they required.

Only now is man realizing his mistake.

Of course man has paid an extremely high price for his defiance of Nature. He had paid—is paying and will continue to pay in ill health for his inability to understand those common processes of nature with which he has so ruthlessly interfered.

Civilized man looks with disdain upon the savage in his dingy hut—his lack of sanitation—his failure to see the advantages of the radio, the ice box and the fireless cooker.

Civilized man believes that his steam-heated apartment—his running water—his air conditioned rooms—are far superior from a health standpoint than the rude, insect-infested huts of the aborigines. Yet civilized man has tens of thousands of ailments—disabling ailments—that fill books as large as an unabridged dictionary. On the other hand, the savage—with all his squalor—could probably count his maladies on the fingers of his two hands. Not only that but the savage is free from those breathtaking diseases of cancer, diabetes and a host of other common, ordinary dangers that civilized health authorities cannot seem to subdue.

Healthy Natives Live Long And Contented

Right in our own country we can find examples of good health that should make civilized man ashamed of his transgressions. In the so-called fever ridden region known as the Everglades of Florida—those sub-tropic wastes of reed-grown creeks and bayous—live a tribe of Indians known as the Seminoles. These Indians have steadfastly refused to adopt the white-man's way of life.

Today in their primitive state they still remain healthy and robust. Babies are born without fuss and bother. No attendant nurse and bespectacled doctor are there to give sage advice to the mother. She is usually unattended and brings forth the youngster in solitude—without assistance of any kind. Her diet is a natural one. It is procured from the native soil—unsullied by artificial fertilizers.

The Seminole mother has no refrigerators—not hot water bags—no blood transfusions.

Scientists grudgingly agree—judged by the fate of other Indians—that if the Seminoles had become civilized—had adopted so-called civilized improvements, they would not have survived. Their sole protection has been and is that they dislike the white man—and refuse to imitate him or adopt his way of living. Yet the Everglades abound in all manner of insects, pests and dangerous snakes. But the simple natives are able to protect themselves from natural enemies.

Even the bite of a rattlesnake—the scourge of the Everglades—has little effect as far as these savages are concerned. These Indians possess herbs to counteract this reptile poison in a most successful manner. Such herbs are Nature's medicines. They are the product of no laboratory and possess no pedigree of scientific research to make them "popular."

The Seminoles as a race are happy in their freedom. Their lives do not know the restraint of other people or the physical restraint of ailments. These savages need no drugs to enable them to sleep—no purgatives to free clogged up bodies from the poisons generated by unwholesome food.

In contrast with this seeming immunity to ailments of the Seminoles civilized man certainly has not been fortunate.

Early Man Was Powerfully Built

In the early eras civilized man was well nourished and well built. With the advance of his culture—his learning—his health failed him—retrograded. He went back physically. And much of his unfortunate condition is due to his abuse of the soil that gives him his food. Ruthlessly he disorganized and disturbed the balance of Nature—the balance that she had been slowly establishing through countless ages! His fields are worn out—his food is lacking in nourishment and the worms and other pests are getting beyond control. He has tried to make up for his abuse of Nature by the most technical means. He uses heat, ice, preservatives and canning to prevent his foods from deterioration. Every ingenuity of an intelligent mind is brought into play to give him better and more healthful food. But he does not face the situation squarely. He seeks to ameliorate present conditions without attempting to rectify the underlying cause which is the true basis of all his food troubles. And today's burning question is: How is man succeding in his efforts? The frank answer—the only answer—to that question is: "Man is failing miserably!"

Dwindling Soil Nutrients Cause Hidden Hunger

The human race is already on a starvation diet. Deficiency diseases—caused by the lack of needed minerals and vitamins—admittedly are on the increase. The future is anything but reassuring.

In many parts of the world and in the United States in particular there are many places where the soil has been so mistreated by man that the food grown thereon is positively dangerous to health. Not only that, but in every part of the United States there is a decided dietary lack of two major minerals needed by the body. These two minerals are calcium and phosphorus. Calcium makes up the major part of our bones and teeth. Phosphorus feeds our brain and nerves.

This lack of calcium and phosphorus is doubly strange in light of the fact that most of the commercial fertilizers are rich in calcium and phosphorus. Yet the soils where these ordinary commercial fertilizers—rich in calcium and phosphorus—are used most abundantly are actually producing foods which are deficient in these very minerals.

For example, such deficiencies are becoming apparent in northern New England where even the presence of ample quantities of Vitamin D cannot prevent the spread of rickets. In Florida there are other mineral inadequacies, particularly of copper, zinc and manganese. In some sections of the country there are also shortages of lime, iron, sulphur, magnesium, phosphous, boron, iodine, and potash.

Yet in many of these affected sections there has been a copious use of ordinary commercial fertilizers which consist mainly of nitrogen, phosphorus and potash.

When the scientist Leibig perfected a means to make a concentrated chemical compound of minerals to be used as a fertilizer, he was hailed in many quarters as a Messiah. Yet on farms, consisting of light soils, the continuous use of such chemical fertilizers often decreases rather than increases its productiveness over a period of years.

While these fertilizers do give to the soil a number of elements the plants can use—however many of these fertilizers are deficient in some of the rarer minerals and other elements the plants need for proper growth. Again, ordinary chemical fertilizers when applied to land which is deficient in organic matter, seep through the soil and a large proportion is no longer available in the form in which it was applied and some is even carried away in solution by the rain—thus becoming a complete loss.

Yet it is a foregone conclusion that if the soil is robbed through the harvesting of crops—that which is taken away must be returned in some manner.

In one section of Florida I noted a well situated farm that is worn out. Thousands of dollars worth of chemical fertilizer failed to save its productiveness. Yet just adjacent to these denuded acres are natural jungles growing luxuriant vegetation. These jungles have received no chemical fertilizers—no human help—and I may add—no human interference. Nature has supplied everything necessary for all the needs of this most beautiful vegetation. Was there ever a more astounding contrast, than this?

Not only is there a lack in our soil of the necessary elements to permit the plants to grow properly, but farmers report unanimously that they are forced to use more insecticide than in the past. Bugs and worms seem to be progressing as civilization advances. Bugs and worms are becoming immune to the simple insecticides that worked well in the past. Each succeeding insect generation seems to be stronger than the last.

The manufacturers of the insecticides must increase the strength of their products to make them again effective. Since some of these poisons which make up the potency of the insecticides remain on the crop after it is gathered, automatically there has appeared a new menace to human health. Now this menace to health is not one of passing fancy. There are numerous cases of actual poisonings of human beings from the use of fruit and vegetables from which the insecticides had not been removed. Luckily regulations have now practically overcome this menace.

Well-Fed Plants Have Natural Protection

A healthy plant has natural protective agencies which keep it in a thrifty condition.

These protective agencies come in many forms according to the needs of such plants.

The healthy potato has a chemical element in its skin called solarnin. This is a poison to the insects that would feed upon it. And also a rich supply of potash makes a hard skin unpalatable to bugs.

The healthy apple has wax on the skin which turns away bacteria and fungus—microbes. The green apple has large amounts of malic acid which is a protection—a great protection even against youngsters. These acids are turned into sugars as the apples get ripe.

A similar protection is found in healthy oranges, grapefruit and lemons. Here it is in the shape of oil in the skin and citric acid in the fruit, also in its vitamins and minerals.

Also a well-balanced mineral supply in many fruits is a protective feature against bacteria and fungus.

In other words, the plant possessing an ample supply of the proper nutrients can provide itself with natural insecticides in the shape of sugars, gums, waxes, resins and acid or alkali and or alkali compounds of various kinds. Of course, if a plant cannot receive from soil these essential elements it cannot protect itself from the ravages of such pests.

A healthy plant—a well nourished plant—will be less affected by worms, bugs, insects and other pests than will an undernourished plant.

Just as the insects and worms are becoming more hardy, plants which furnish us food are likewise undergoing a change. The soil on which they grow is so universally deficient in many things the plants used in the past for proper nourishment that it is not remarkable that these plants have accustomed themselves to short rations. You see, Nature does not give up easily. Nature fights stubbornly to bring forth her vegetation. But Nature cannot ignore facts. If the plants cannot find enough water in one place—she changes the forms of the plants and brings forth cacti. If the ground is too moist—she makes great branching roots with air vents to support the heavy tops in the soft soil—and we have the cypress. These vents are in what is termed the cypress knee. And when she found that the farms of the civilized world were lacking the nourishment that should be there, Nature had no hesitancy in changing the structure of the plants to meet this meagre existence.

Gradually Nature has been constructing simplified plants for our denuded farms. These plants require less and less of certain minerals. These plants have the same outward shape but Nature often substitutes other minerals in the place of those minerals she used so effectively in the past.

Plant Appearance Deceptive

The plants that are results of this lack of minerals may look approximately the same as the plants of yore. But they may contain minerals that are actually harmful—instead of healthful—to the human body. That is something scientists seldom think about. And yet it is a condition that is already upon us.

Plants When Starving Substitute other Elements, Some-times Poison Such as Selenium

For example, if Nature finds that lime or potash is lacking in the soil—or if present in the soil in a form Nature cannot use—Nature may substitute, as food for the plant the minerals magnesium and soda in the place of this lime or potash. Or she may substitute in the place of sulphur, selenium—an element which may exert many harmful conditions in the human body. Persons eating food containing selenium are subject to blindness, loss of hair and grave nervous disturbances. Several cases have been reported in the Dakotas, even blindness, loss of hair and death—making land worthless.

Soil Depletion Now Widespread

This process of change by Nature is not confined to one locality. Lately it has been discovered in many sections of the country. One particular change has received much publicity. It has to do with a common weed known colloquially as horse tail. This weed has been gradually changing in structure so that now scientists have discovered that one of the minerals Nature has injected in the plant is ordinary gold. Yes, horse tail has become a gold collector and where there is gold in the soil—and a lack of other essential minerals—the common horse tail weed becomes a miner.

Now let us hope that this does not spread to the garden plants. Gold is all right in your purse but taken internally—in a carrot for instance—ordinary gold will hardly help nourish your body and might conceivably do harm.

Of course not all the substitutions of minerals in plants made by Nature are harmful. Yet many of them are.

Disease Traced to Soil Defects

Also, scientists have discovered, that when the soil is deficient in some particular mineral—and there are no others to substitute—the plants will grow to maturity bearing only a fraction of that mineral it should have given in abundance. Calcium is often deficient in many of our modern garden plants. And when the food we eat lacks enough calcium we may suffer from various respiratory diseases such as pneumonia, bronchitis, sinus troubles. When calcium is added to animal feed it prolongs their lives.

We, on our part, are apt to blame a germ for these conditions. The truth is that the germ would probably be rendered harmless if there was enough calcium in our body.

A great many persons do not understand how a disease germ gains a foothold in the human body.

As an example—the germ of tuberculosis uses an acid to destroy a small spot in the human body in order to get a proper place to start, as it lives on destroyed tissue.

When there is plenty of calcium in that tissue on which the germ settles—when the human body is well stocked with this essential mineral—not only in the bones and teeth but in the blood and every cell, this calcium—being lime—will neutralize the acid used by the germ. Therefore the colony the germ would start is stopped before it really gets started. This neutralized combination of germ acid and calcium is called formate of lime. There is no doubt that extra calcium in the body helps to fight—and conquer—tuberculosis.

Human beings are alkaline creatures. Their live bodies are slightly more alkaline than neutral. So their food should supply continuously plenty of alkaline materials—minerals and vitamins—in order that their blood stream may be able to carry out the work Nature assigned to it. If the blood stream should become even slightly acid—the human body would die. In fact, when a person dies—his body is acid and negative.

Yet in all activities of the human body acids are formed. So a reserve supply—an extra amount—of alkalis must always be present in the body to neutralize these constantly forming acids.

Heat Energy Turns to Acid

When you move your arm—some of the energy sugar in the cells of your arm muscles is turned to acid. The blood instantly neutralizes this acid. If, through exertion, the supply of acid is built up faster than the blood can neutralize it—the human body becomes tired. Exhaustion is nothing more than an excess of acid or waste matter in the muscles or system. As the mineral reserves are used up the blood begins to rob the tissues in the body to maintain its essential balance to the detriment of such tissue.

When the blood is well-balanced—when it receives from the food we eat everything it needs for its work in the body—we in turn enjoy most vigorous health. There is an old saying that you are only as well as your blood is well.

To make this well-balanced condition of the blood, the

body must receive in the food that is eaten everything it needs for properly repairing the worn parts of the body. And this nourishment must come from the air we breathe—from the water we drink—and from the food we eat. So soil that is capable of raising proper food is of paramount importance to our health.

It has been found that the soil—on which most of our food is raised—has been so much depleted in minerals that the food grown thereon does not supply the body with the nourishment it should have. Therefore we have arrived at a place where the adding of additional minerals and vitamins to our actual food has become necessary to reduce mortality.

Reinforced Animal Feeds Improved Health

In a carefully conducted feed test with baby chicks those supplied with balanced minerals had a 75% lower mortality rate.

In another test with valuable silver foxes the ones fed with balanced minerals were not so nervous—did not eat their offspring and the fur had a finer texture and color—the pelts brought higher prices.

Some sick Guernsey cattle suffering with Bangs disease garget and Mammitis made a hasty recovery when the minerals were supplied to their feed, and when fed to milking cows the quality and color of the milk improved.

Hogs with travel sickness soon recovered when given the minerals and their mortality rate was greatly reduced.

Human beings—given the same minerals, in a refined form—have obtained outstanding benefits from them. Since this was written scientists have claimed cures of diseases.

There are hundreds of other cases over a wide period of time that could be cited. This leads me to believe that what happened with the chicks, the foxes, the cows, the hogs and human beings is but a natural consequence of right feeding. Just give Nature what she needs—give her the proper material for her work—and disease and ailments will be greatly reduced. Science now confirms this quality. See later.

During the past fifteen years it has been demonstrated in a thousand ways that human beings and animals are not getting the minerals their bodies need for health. Also it has been demonstrated that the addition of minerals to the food—providing those minerals are "live" minerals—

organic minerals—often will solve many questions of ill health. But adding minerals to the diet of either animals or human beings is a make-shift kind of remedy. It is a temporary measure without in any way correcting the cause of this mineral shortage.

Nature Builds Soil — Soil Can Now be Revitalized

The situation must be approached from another angle altogether. It must be approached through the revitalizing of the soil—through a closer co-operation with the methods used by Nature.

Conservation is needed—but not the conservation in the general sense as we understand it today. It is more than a mere rotation of crops—the growing of one crop one year on a given plot and another crop the next. This rotation does some good but as each succeeding crop takes from the soil some of its richness—each successive crop, as now handled, robs the land.

How Soil Problems Were Solved

It is time for a new deal for the farms. It is time attention was given to just why soil produces crops. Until that is done slow starvation will continue to creep upon this nation—yes, and upon every nation of the world. It is time for the farmer to stop playing blindman's buff when it comes to planting crops on his farm. It is time that he knew—in advance—just what plants his particular soil will produce. He must know the plants most adaptable to the reactions of his land. Some plants thrive in an alkaline soil—many in a neutral soil and some in an acid soil. A plant growing under an unfavorable reaction will not thrive, so the grower should plant the type of crop suitable to the reaction and type of soil to get good results.

I perfected a simple soil tester for this purpose. By its use the soil can be analyzed in about one minute. A simple chart indicates the variety of crops suitable for such soil. It also may be used to show what changes should be made in the soil to make it suitable for any desired crop. For this purpose I designed a kinetic analyzing machine which became invaluable in the detection of trace elements in soil, before I was able to obtain a spectroscope.

So the first move for real food improvement in this country is to take the guessing out of the sowing of the seed.

The next move is to put back into the soil those things that the crops have already taken from it. And this return to true fertility is a question of the use of proper plant food.

Back in 1900, the eminent Professor Crooks made the positive statement that this continuous robbing of the soil by man would result in an absolute shortage of nitrogen in the ground. There was but one source of supply as far as organic nitrogen compounds were concerned. This supply came from the islands off the Chilean coast of South America. And this supply was being rapidly depleted.

For tens of thousands of years the droppings of countless seabirds had been storing this organic nitrogen-filled fertilizer on these dry, barren rocks. But with the huge demands of the tens of thousands of hungry farms throughout the world the available supply of this bird-made nitrates was fast disappearing. When it was gone—as it would be within fifty years—Professor Crooks could see nothing for civilization than a slow starvation from a lack of food containing the proper elements needed by the body.

Cheap Nitrogen Can Now Be Had From Bacteria In "Organo"

But Professor Crooks was needlessly pessimistic. Invention and cheap water power made the mechanical fixation of nitrogen from air a commercial possibility. So nitrogen shortage was avoided. The huge investment of the United States Government in the Tennessee Valley is partly the outgrowth of Professor Crooks' warnings. Soon the American farmers—as well as the explosive manufacturer—will have an inexhaustible supply of nitrogen. The farmer will receive his in various compounds as fertilizer and will thus be independent of the Chilean sea-bird fertilizer.

Man is trying to do mechanically—electrically—what Nature is only too willing to do naturally. This earth existed triumphantly for tens of millions of years before a single pound of bird droppings was brought from Chile and before a cubic foot of air was forced into the compressors at a nitrate plant. And Nature has not abandoned us. She is always ready to take the nitrogen from air and give it to the roots of the plants as bountifully as she ever did before. All she asks is that we do not interfere with her delicate balances as we have done in the past.

It is time to point out that good land is really live land. The ground that will grow fine, healthy nourishing plants is just as much alive as the plants themselves. Live land is filled with microbes, organic matter and minerals. The ground about the roots of the plant must simply teem with microscopic life. If the numbers of these organisms are reduced—if the use of the wrong kind of fertilizer or poor land treatment prevents their multiplying in large enough numbers—the plants that do grow will not be vigorous plants and their value as food will be reduced.

Of course, back in 1900 when Professor Crooks made his doleful prediction about the coming shortage of nitrate fertilizers, few persons realized that there were bacteria in soil which had the power to extract nitrogen from the air. Some people believed that nitrogen was taken directly from the air by plants but that was not the case. This is only accomplished by certain groups of bacteria.

Legume Bacteria Gather Nitrogen

In 1905 I made pure bacterial cultures which had the peculiar ability of taking this nitrogen from the air and storing it in nodules in the roots of growing legume plants. I distributed this nitrogen-storing bacteria throughout the United States. Several prominent authorities doubted the value of such bacteria for such work at that time. Two previous attempts had been made prior to mine without success. Mine certainly was a voice in the wilderness. Farmers knew little—and cared less—about the so-called beneficial microbes or bacteria in their soil. The only microbes they knew about were the kinds they sought to exterminate—the harmful bacteria that destroyed their crops. For a germ or a bacteria to be helpful to plants—well, that, to the average farmer, seemed impossible, pure bunk.

Despite this almost universal unbelieving attitude on the part of the tillers of the soil, I was able to introduce these pure cultures of bacteria in various parts of the country. Their success was outstanding. They so improved the crops where they were used that repeat orders soon spread the business from coast to coast.

In a few years, because of the increasing demand for such bacteria cultures, practically every seed dealer in the country recommended them. These bacteria cultures became one of the most profitable means of fertilization for the enrichment of land that the farmers had ever used.

That was many years ago. Today, no sensible farmer would fail to inoculate his seed with such nitrogen-storing bacteria. The farmer has found that the adding of this bacteria to the seeds he sows, many times made the difference between the success and the failure of a crop.

Of course seed inoculation with bacteria had to go through its trial by fire. Many eminent agricultural authorities condemned its use. It was not until the results were so outstanding that the possibilities of seed inoculation could not be ignored. Then these same authorities who had so vigorously condemned the practice acknowledged that seed inoculation was in fact effective. Yet even after it was definitely established that the use of pure nitrogen-storing cultures of bacteria for the inoculation of the seeds was the proper method of getting such bacteria directly at the spot where the work of growth was to begin, many authorities tried to substitute plain soil known to be impregnated with such bacteria. The substitution did not work out satisfactorily. While this substitute bacteria-impregnated soil held bacteria which stored nitrogen—it sometimes held plenty of harmful bacteria that did great damage to the seeds and later growth of the plants. So after trial—and many disappointments—this bacteria-impregnated soil was abandoned for pure culture inoculation of seeds.

Now to get nitrogen from the air into the soil—at the roots of the plants where it can be utilized—billions upon billions of these tiny, natural factories must work in every acre of land. These factories exist in the form of bacteria—little living organisms that have just as much life in their particular sphere of influence as man has in his. These bacteria can subsist only under proper conditions. Without these conditions they die.

Some of these little organisms fix or gather the gaseous nitrogen of the air—and feed it to the roots of the plants. That is their sole purpose in the scheme of things that Nature had devised.

Man did not realize that the field he called fertile was simply a field so constituted that these helpful little organisms—these bacteria—could thrive and prosper.

If you were to take a portion of rich loam and put it under the microscope you could discern billions upon billions of little moving creatures. The decaying vegetable matter—which makes up good loam—might be likened to a cake of yeast which is a mere square of billions of tiny live organisms.

A few years ago I was invited to Miami, Florida, to speak before a gathering of leading citrus growers of that state. For months I had been traveling through the various counties of Florida. I had seen the many lime trees withering and dying. I had seen thousands of acres of grapefruit trees struggling for a bare existence. I had seen untold numbers of oranges that had prematurely dropped to the ground without ripening. All this because the ground on which they grew had failed to furnish the nourishment they required.

A Tropical Legume Will Aid All Hot Soils

For the most part Florida consists of white sand over which is a thin film of loam which soon becomes exhausted.

Now on this particular day when I was driving through southern Florida on my way to the meeting in Miami, I made a tremendous discovery. The passing glimpse of a green spot of vegetation made me stop my car and walk back along the road for some distance.

If I had seen an angel in that out of the way place, I could not have been more surprised.

My companions could not understand my sudden interest. They were anxious to be on their way. Already we were late for lunch and they had no desire that I should keep them waiting. But as I walked slowly toward the car—carrying in my hand a small bunch of green clover—the importance of that meeting of prominent citrus growers in Miami had taken on a much more important aspect.

My companions smiled at my enthusiasm over this clover. It had been brought to Florida several years before along with numerous other seeds. That it had survived and was flourishing—had meant nothing. But to me it was tremendously important. I felt like a modern Ponce de Leon who had made a far more interesting discovery than a mere fountain of eternal youth. Because on the roots of this fine leaf clover clung little nodules full of nitrogen bacteria—a veritable storehouse of atmospheric nitrogen compounds. Yes, this vigorous clover assured the future fertility of this great state—and a wonderful fodder crop as well. Its fine stems and many leaves made an ideal green manure crop for the land.

No wonder that afternoon, when I addressed the members of the Florida citrus exchange in Miami, my prepared talk veered to the great value that lay in this marvelous clover. Here was the source of billions of Nature's little nitrogen factories that held in them the secret of unbounded fertility. Here was the plant for which Florida agriculturists had been searching for generations, a clover that would grow in Florida heat.

My extemporaneous talk as to the possibilities surrounding the future of this clover did not fall on deaf ears. The orange and grapefruit growers did not realize then that in that tiny bunch of clover I exhibited to the meeting was probably the solution to many of their difficulties but their interest was aroused. Such innovations take time.

However, today you find that the benefits of this clover are fully recognized. Not an issue of any agricultural journal in Florida—and other southern sections—but had scores of concerns calling attention to the marvelous, soil-building possibilities of Alyce clover. And this plant—a legume—is just one of Nature's methods of extraction of nitrogen from the air. Just one of Nature's methods that are far more efficient—far more complete and far cheaper than the best type of mechanical apparatus ever invented by man.

A good soil must have many different kinds of helpful bacteria—the proper organic matter to keep these bacteria alive and various different kinds of minerals on which these bacteria can operate.

Variable Qualities of Organic Compost

It has long been known that ordinary stable manure helps make soil rich. Stable manure gives to that soil far greater fertility or growing power than the so-called recognized mineral fertilizers which only contain nitrogen, phosphorus and potash.

The reason for this is not difficult to understand. As stable manure decays it sets free many different kinds of acids which dissolve the minerals in the soil. Also decaying stable manure provides excellent food for many forms of bacteria. These bacteria in turn produce enzymes and acids which further make the many minerals in the soil available for plant food.

These manure acids and enzymes together transform the

minerals—nitrogen, phosphorus and potash—into substances the plants can use as food.

That is why organic—or alive—fertilizer is needed to make these minerals available to the plants. If the land cannot grow these bacteria, no amount of chemical fertilizer will permanently make the ground produce profitable crops. One might say that the fertility of a piece of land is strictly in proportion to the amount of bacteria that will grow on that land. Chemical fertilizer must be accompanied by live organic fertilizer if the soil is to be made truly productive.

Types of Soil And Microbes

Now soils are of three varieties:

First—those soils plainly on the alkaline side which will produce certain definite crops carrying a predominance of alkali minerals.

Second—neutral soils—neither alkaline nor acid—which will support a vast variety of plants.

Third—acid soils which are adaptable only for a limited number of acid vegetable growths.

There are bacteria—Nature's little workers—that live comfortably in all of these three different types of soil. It is the duty of the farmer to see that they continue to thrive. So long as the grower keeps the bacteria in his soil alive and in abundance, his crops are going to pay big dividends in health and money.

If soil continues to receive quantities of only chemical fertilizers and no additional organic substances to feed the bacteria—the natural balance of the soil become deranged and the proportion of organic matter suitable for food of these bacteria is depleted. Thus the bacteria do not find sufficient food and do not increase in sufficient numbers. In fact, many of them die of starvation and although great quantities of chemical fertilizer may be used, the ground produces less and less each year and becomes unprofitable.

Poor Farms Harbor Poor Microbes

Worn out farms—as a general thing—are due to lack of understanding. Few farmers appreciate the need of bacteria for plant life. Many fields with enough minerals to produce bountiful crops for a thousand years, are barren today merely because there is no suitable food in those fields to feed the bacteria. This is not supposition. It has been proved in many, many cases.

We are removing from the soil of our farms its vegetation for market. Most of this vegetation is discarded as garbage and goes to the incinerator or is dumped into the sea. It is lost forever. Thus we are draining our soil of its bacteria-feeding substances. We foolishly believe that we can compensate our farms for this loss by supplying quantities of acid mineral fertilizers. But this is a false conception. Not only are many of our farms—because of the lack of bacteria—unable to digest this acid mineral fertilizer and sadly deficient themselves in many minerals the plants require for their proper sustenance. Particularly is there an absence of copper, iron, molybdenum, zinc, manganese, boron and sulphur. Thus the foods that finally struggle into existence on such half-fed farms often lack many of the essential minerals needed by the human body for its growth and health.

Losses of Organic Matter In Soil

All the minerals a human being needs each day to sustain health could be placed in a teaspoon. Yet it takes approximately two pounds of food to furnish an average human being with these essential nourishing elements. Scientists have proved that in order to place two pounds of miscellaneous food in the stomach of a human being, at least ten pounds of plants must come from the field, the garden or the grove. The wastage is enormous.

The farmer first trims off the outer leaves—then the commission man on his part does some trimming—followed by the knife of the retailer and finally the ministrations of the cook. And to place two pounds of food in the human stomach ten pounds of plants are taken from the farm. It is wastage pure and simple. Most of this ten pounds of plants go into the garbage—the rest disappears down the sewer. The part retained by the human being could nestle easily in the bottom of a teaspoon. And the farm, on which these ten pounds of plants were raised, has lost forever the substances they contain. And yet, these same ten pounds of plants—that are so ruthlessly thrown away—would support enough bacteria—with the proper proportions of mixed minerals—to bring forth a full acre of splendid crops. That is enough food to support a human being for an entire year!

A Balanced Soil Needs Little Help

It is said that a good soil should never wear out. The more the growth, the more it should produce. That is the natural way. This of course is dependent upon the balance of the organic food of the bacteria. Just remember, the rocky particles—the minerals to be acted upon by the bacteria—you might say—go down to China. In a good soil the supply of these minerals is well-nigh inexhaustible. Of course, climate, water and the acid condition of the original soil play important parts in what any given land will grow. Crops must be adaptable to the environment.

I have seen actual waste lands, composed mostly of sand, built into a permanently fertile condition by proper handling. One farm of this character is in New York State. It was badly managed. The owner tried to raise grain—in other words, he tried to make it produce carbohydrates when the soil showed—by test—that its crop should have been of a protein nature—such as beans, alfalfa, etc.

Naturally, the whole enterprise was being run at a loss. Man cannot successfully buck Nature. So the introduction of additional organic substances to support the right kind of bacteria for a protein crop was made. The crops were changed. In a single season this farm became unusually profitable.

How much better it is—how much more sensible it is—to give to the ground, in return for the food that is taken away—organic substances that will feed the life-giving bacteria in that ground. And giving to the ground this common-sense organic fertilizer is not difficult as you will see. This has been demonstrated in many parts of the world.

But the public must be awakened to the urgent need of doing this. The public must be willing to co-operate. The public must realize that if it is to eat good food, the bacteria in the soil of the farms that raise this food must have satisfactory and plentiful nourishment. If this organic nourishment is supplied to the bacteria in the soil the plants produced on that soil will be loaded with minerals and vitamins and will contain the essences which give color, odor—or I might say perfume—to the food material.

Food Shows Greater Nutritional Value With Organo

It is claimed that analysis shows a higher mineral and vitamin content of oranges so treated. This should be the case, because this organic mixture carries all the various elements required for growth in a well-balanced form; thus the plant is better nourished. It has been thought that the rare elements could be applied in the regular chemical fertilizer, but they seem not to produce this same satisfactory result as when nature combines them together in the organic form, and some—like iron, boron, manganese, zinc and copper—are required in very delicate quantities. In the organic products they are just as Nature produces them so naturally they are in the most satisfactory quantities to meet the exacting needs of the growing plant.

Organo Saved Ailing Trees

One large lime grower in Florida—in fact the largest in that state—reports that his entire grove was threatened with extinction through the appearance of a disease called "gumosa." Now gumosa merely means that for some unaccountable reason the bark of the lime trees had split and the sap was running out instead of being carried to the leaves and fruit. This grower was advised by agricultural experts that his case was hopeless—they suggested that he cut down his grove. This involved a great loss of money and time. So he looked about him for other counsel. Finally he was convinced of the advisability of trying this organic type of fertilizer made from waste organic material. The bacteria found ample food in this organic fertilizer and increased in great numbers. The dead minerals in the soil were converted into food that the lime trees could use. And the trees, thus strengthened by additions of minerals that they had formerly lacked, fought the ailment gumosa—and won. The wounds in the bark closed and the trees were saved. It would seem that this transformation was accomplished merely by supplying a "balanced" diet to the trees that had heretofore been existing on a starvation diet.

Fruits Have More Food Value

Oranges, grown with this type of organic fertilizer, are far superior to those where the trees are given nothing but chemicals. A close check for over two years showed that these oranges, grown with organic fertilizer, have a superior color, ripen better, are juicier and sweeter, travel better and have more vitamins and minerals.

In fact in one particular grove not far from Orlando, Florida, where the citrus grower specializes with tangerines, the effect of the addition of this organic fertilizer was outstanding. The tangerines grown were far different from any other tangerines grown in the state. Instead of the typical tangerine—a small shrunken fruit in a wrinkled skin—these tangerines were full juiced. Their aroma—their tang—and their deliciousness of flavor made them superior in every way.

This new type of tangerines found an immediate local market. They never left the State of Florida. All were purchased and eaten right in the neighboring city of Orlando—because those folks living among the oranges, know good fruit when they see and taste it. These new style tangerines—the result merely of the addition of the organic fertilizer I have described—are now called "Champaigne" tangerines. And it is a name they richly deserve. Not only is the fruit better but the yield of the trees—without a short three years—has been doubled.

Now this organic fertilizer did not make a new kind of tangerine. Not at all. This new kind of organic fertilizer gave Nature what she needed for the first time in generations. The tangerine as it emerged in this particular grove was the tangerine Nature intended—the little, wrinkled skinned, half filled tangerine of the market is a starved type of fruit that shows only too plainly its lack of nourishment.

Celery raised with this type of organic fertilizer is much better than that grown ordinarily. This celery—reinforced with many minerals—does not wilt as quickly as its underfed brother. Not only has the organic fertilizer grown celery more hardy but it has less fibre than ordinary celery.

In the State of New Jersey, a large field devoted to the raising of broccoli was divided exactly in half. On one side was used the ordinary fertilizer of commerce—the best money could buy—and the other side received only the organic fertilizer I have described. The result in a single season was outstanding. The broccoli grown on the usual type of fertilizer were ordinary—run of the mill broccoli while the broccoli grown on the other half—the organic fertilizer broccoli had larger heads, were better to the taste and brought advanced prices. Incidentally, there was far less disease among the broccoli that received organic nourishment.

In Delaware we find better peaches. On Long Island we find better cucumbers. In Connecticut it is flowers—carnations and roses—stronger, larger and longer stemmed. From far away China it is rice and vegetables. From Africa it is corn and oats. From Manila it is sugarcane with stems 2 3/8 inches in diameter against ordinary cane stems of 1 1/2 inches.

Prospects For Better Food Reassuring

Now, what does the future hold for the people of the United States? Is it a slow starvation for lack of minerals as visualized by Professor Crooks back in 1900? Are the people of America to gradually disintegrate, losing teeth, hearing, hair and ultimately life itself? Frankly, I do not think so. The picture, as I see it, is anything but a dismal one. I look forward with confidence to vastly improved conditions. It is not too late. Nature has not deserted us. On all sides I find a much greater willingness on the part of science to investigate the fundamentals of plant growth. I find a desire on the part of agriculturists generally to get a better understanding of this industry we call farming.

The farmer of the future will raise crops scientifically that are ideally suited to his land—the climate and the demands of the market.

Guessing will be eliminated wherever possible. The bacteria in the soil will be just as carefully guarded as are the stables, the pens and the stys. The farmers will give back to his soil the minerals of which it has been robbed. The plants will be healthier plants, better able to fight their own battles. All this I believed in 1941, then came war causing a ten years delay in my life's work.

A Spectroscope Showed Scientists Cause of Disease

I didn't expect such luck that there would be a moment's discovery by scientists of the Missouri University to confirm the value of minerals as a means of overcoming some diseases of animals and human beings. Their research was more complete than had been the case in my work, but was discovered by them in a somewhat similar way by analyzing soil as I had done, and also the blood of subjects by use of a spectrograph. By this means they were able to discover the cause of the disease. The blood showed the missing elements which had caused the disease, and when they supplied the patients with the missing factors all but few made slow but complete recoveries, even in chronic diseases, which had failed to respond to the sulfa drugs or to antibiotics. They found that cattle were diseased through malnutrition due to living on soil depleted in trace minerals. They then returned the missing minerals to the soil, and to their surprise the cattle sought out such treated soil and made recovery of their diseases by their own instinct for survival. Other animals were given the minerals in their food and it worked equally well. They all recovered from their disease. This can all be verified by reports in bulletins of the Missouri University, Johns Hopkins University, Merck & Company, Drs. Albrecht and Allison, and other equally famous co-workers, leaving no doubt of this greatest of all discoveries. It means a future of improved living for all mankind. It removes the mysterious of many baffling diseases soon to be mastered by simple analysis of soil and blood, as I had reported with rodents many years ago.

I Keep Secret Reduction In Diseases of Plants And Animals Recorded

Later Sir Albert Howard began without fear to claim freedom of disease in cattle with his work on ordinary compost. I then felt more free to speak of that prospect. Later the classical work of the Missouri University and other scientists showed that when soils were supplied with completed plant food, animal diseases were cured, and went even farther and showed that the blood of animals and humans living on food from deficient soil acquired diseases of several kinds, and when these missing elements were given to such patients all received improved health.

Now Scientists Say Improved Soil Cures Diseases Just As Trace Minerals Do

I now feel it safe to bring forth in these pages the memoranda I had written a quarter of a century ago with the hope that it will add some impetus to this important development. When I carried out my research I did not have the valuable spectrograph to aid these findings in analysis, but I was fortunate in being able to make a device which I called a kinetic analyzing machine, which worked well, until I was able to purchase one of the early spectroscopic microscopes which was more difficult to use but confirmed my work with the kinetic analyzer. It has shown me all the various trace elements in soil and their amounts.

Kinetic Analysis Gave Me The Cue to Overcome Maladies and Saved My Life

Dr. Cameron, of the U.S. Department of Soils in Washington, in 1910, hearing of my kinetic machine called on me to see it operate as he said that he was surprised at some of my soil work where I would recommend to farmers what to use and how much. He said that he had nothing like it. I told him that we were a small company and had kept it a trade secret for obvious reasons, but that as he had come from Washington I would show him how it worked, provided that he would not disclose it to anyone. He agreed that it would be kept secret. I took him to the laboratory where six girls were testing a number of soils for farmers. Such soils came in daily. We had a printed sheet and marked the elements that the soils needed to produce good healthy crops so that the farmers would provide only the elements needed. Dr. Cameron was very much surprised, and said that he was sorry that he could not take it back to Washington, as they had nothing like it there. He assured me that he would not disclose it until I was ready to do so. I used this analyzer to show Mr. Jeffers, of Walker Gordon Company in 1908, how to build up their farm at Plainsboro. The land was in very poor condition, due to a deficiency of a few elements. The alfalfa was moth-eaten with crown rot and little of it left. It took me six hours to convince Jeffers that I could help him. Only then he would try but four acres with the new treatment because the New Jersey Experimental Station had told him the land was unfit to grow alfalfa. Within two years he was producing the finest alfalfa in this country, and he became one of my best customers. Mr. Walker, the owner, told me the farm had been valued at $40 an acre, but had never been worth it, but that since I had built up the land it was worth $400 in production per acre.

He wanted me to develop all lands in New Jersey that way as his farm in Massachusetts had also shown equally good results. He said that all soils in New Jersey could be improved to be worth $400 per acre. That farm is famous today for its production record and the high quality of its milk. I never told Mr. Jeffers how it was done. The same applies to the Gonzales Farm in South Carolina; Fallens Farm there; and Gov. Haywood's too. The three bales of cotton I produced per acre for Boylson, and the two million dollars increase I made for Liberty Farms, Sacramento, within two years from a defunct farm. Of course, there were many others, but these are the best examples. All the result of use of my secret analyzer.

I Used Kinetic Analysis to Win Friends

My advantage was great, chemical analysis in such cases was a waste of time and fooled everyone, but as Missouri University has recently demonstrated the life of soil and those living on it depends entirely on kinetic analysis or better still spectrographic analysis to show how sick it is and those living on it. It will take about ten years more before this work is fully evaluated. This despite bulletins by able men giving full disclosures of this greatest of all discoveries. There are still skeptics of this who believe that you can build a sound body without the need of the essential elements it requires for perfect health. When food lacks essential elements we fail in health and need injections of hormones such as Acth, Cortisone, and when soils fail to provide minerals and vitamins, then the doctors make injections of the hormones, vitamins and minerals often with spectacular results, all unnecessary if we restore the soil and get a good balance of all such essentials from our meats, eggs, milk, vegetables, grains and fruits. It doesn't matter which way we obtain such nutrients so long as we obtain them. They all work the same.

Doctors Fail to Realize That Death Follows Deficiencies

We may ask what does all this story mean to mankind. That is a question only time can tell, a question first of veracity. Then the follow through by able workers to carry on the necessary research to evaluate each detail to the logical conclusions. My endeavors were but preliminary, stumbled upon because I devised a kinetic analyzing device which clearly pointed out to me the dwindling supplies of all available essential plant foods. This was before there were other adequate means to do so.

Spectrograph Proved Kinetic Analysis Correct

The spectroscope and spectrograph came later, and fortunately was used by able men in confirming also that soil could no longer supply sufficient essential elements to prevent disease, let alone cure it. Now I can refer you for confirmation of this startling discovery to the Missouri Bulletins on the subject, also other publications confirming these statements. When I was later able to obtain a spectroscope it showed that my kinetic analyzer was correct in its operations. It would show even the minutest quantity of available elements in the soil. It would also show if none was present. The spectroscope was adopted by me not because it was more accurate, but because it gave the answer in less time and work.

Scientists Find Trace Minerals Cure Disease. Dr. Nolfi's Book Shows Cures of Over 90% of Patients, Some Malignant, By Giving Organic Food

If we are right in our conclusions—also the disclosures already published in the Ludlum Steel Journal, the Missouri University bulletin, the Miracle of the Ozarks, Dr. Nolfi's book published in Copenhagen, Denmark, the magazine Normal Agriculture, and the results in overcoming diseases with a new nutrition from a completely good soil as propounded by Drs. Allison and Albrecht and other great scientists—then we can look forward in the not too distant future to less doctors, less hospitals, less drugs, and a happier carefree longer life with more abundant food of better taste and quality.

We Learn That Soil Can Poison You or Make You Well

When it dawned on us, from our soil analysis, that food was no longer providing the necessary sustenance to maintain people in normal well-being we then carried out the necessary research to meet this need.

We prepared a compound of the various trace minerals with a base to produce a normal reaction in the blood. We followed this with the production of a completely organic food raised from rich organic soil, with an analysis showing that it carried all the 32 essential factors, so rich in vitamins, minerals, and hormones, that it can be sold for less cost than similar synthetic supplies. We also selected lactic acid bacteria of the beneficial kind which would inhibit the growth of putrefactive bacteria in the digestive system. All these products were sent to independent laboratories, hospitals, and doctors for checking so that all claims made for them were substantiated by the different researches made by competent scientists. Some of these researches were carried out for over a year as there was so little knowledge at that time of the values of any of this work. Letters from grateful users came in from both this country and abroad, from doctors, hospitals, and patients, all showing that the researches we had carried out when brought into practical use by the professions and others fully substantiated results of our years of careful checking on this important development.

We began making laevo rotary-lactic acid back in 1905, and they have been continuously on the market ever since. At that time we also introduced lactic acid bacteria for improving the quality of butter.

In 1908 we made up the trace mineral compounds with a base to restore to the blood the missing mineral factors, which we had discovered through the deficiencies in soil, and they have been on the market ever since.

In 1910 we also made up a compound of these bacteria with the necessary food to go in silos for the destruction of harmful bacteria there, which also produced the beneficial lactic acids to cure the ensilage in a rapid and satisfactory manner. Dairymen reported that it not only improved the quality of their milk, but that calves which had to be slaughtered for veal could now be kept to improve the dairy herd because of their improved bone structure. We had food raised from rich organic soil thus providing a complete food rich in protein, vitamins and minerals the first chlorophyll product on the market.

We had to devise means of analyzing soil to determine the available elements in it, and were fortunate in discovering the kinetic analyzing machine. Then we had to have a simple device to determine the reaction of soil, and we perfected the Earp-Thomas Soil Tester, which was patented. This led to the production of complete organic fertilizer for the purpose of building up soil to produce complete food so that the various remedies that we had made would no longer be called upon. This is now supplied with the complete organic fertilizer called Organo, also a concentrate with all factors in 100 pounds to treat an acre of ground.

We found that ordinary composting was not satisfactory so we devised the Earp-Thomas continuous flow digestor, which is now an accepted fact, and in operation in various localities. To have it operate by continuous flow we had to find the necessary groups of bacteria that had no toxic qualities to make the best grade compost which, when reinforced, would meet soil needs. We had to have numerous animals for tests, and growing plants in jelly, so that we could inject them with bacteria before we accepted them, as some were toxic both to the plant and to the animal, even though they did have good qualities as far as producing compost, but their toxins in the fertilizers would have been injurious to the growing plant, and to those consuming it.

It is quite recent that scientists began to discuss the shortage of minerals in our soils, and later that it became a topic as a food problem. All thought little of the value of minerals as they assumed that more than sufficient was available in all soils and foods. Now the quality of the proteins in foods are questioned because they have degenerated in progression as the soil declined in good food production. We now give Acth and Cortisone to aid patients with remarkable results. This is only because we can no longer receive minerals and good proteins to make our own Cortisone and other hormones. When crops and farms are sick how can we expect to retain health on such poor provender. It is now well known by authorities that our soils are sick through lack of minerals. Government maps show extensive sections deficient in one or more essential minerals. While they do not show in such maps deficiencies in microbes of valuable kinds, or inadequate proteins, vitamins, and hormones, such deficiences do exist and will be shown. Now we have that information second-hand in analysis of the crops, and their various deficiency diseases, beautifully illustrated in books. There are pictures of crops of many kinds showing gnarled leaves, necrosis, spotted leaves, fallen leaves and fruit, and blanched leaves through chlorosis, and bark disease such as nectria detessima, etc. The expert looking at the pictures can tell, in most cases, what deficient mineral caused the disease, but when many elements are deficient the plant dies, so he has nothing for reference. Yet we are served vegetables from such poor soil often made to grow with substitute food, such as nitrates, sulfates, and phosphates, anything to stimulate growth to make the plant appear palatable, and in that way to disguise its false food value. When we suffer in consequence we receive vitamins, hormones, and now minerals to temporarily recover until a more serious organ goes bad and has to be removed. As soon as our mineral, vitamin and hormone balance is deranged we have some malady. We don't blame it on a sick cow, or our sick vegetables, or our poor milk deficient in calcium and vitamins. We take drugs to disguise our hidden hunger to tide us over acute ill feeling.

An orange may look good yet be more than fifty percent deficient in vitamins and minerals, its mainstay for health. I have pictures of oranges, on one side which decayed under shipment, the analysis of which showed lack of vitamins and minerals so went moldy and rotted. On the next side it showed oranges grown on good soil, the analysis of which showed it was high in minerals and vitamins. This picture was certified and sent to me with the good oranges which did not decay, but finally dried out whole. This is shown in my 76-page book, with many other such results. They are not isolated cases, but may be expected where crops are grown on soil to which a complete organic fertilizer has been applied. The French officials have recently found that this is true as their fruits and vegetable were healthier. They have not yet learned that their families will likewise be healthier until time proves it for them. I published just a few excerpts from letters showing their results in growing vegetables and fruits in the new complete way.

If minerals did no more than give good health to vegetation, make food taste better, and resist diseases, then they would be of value, but as I have shown, they are life-giving and essential to our well-being. We are indeed fortunate that there is now a cheap practical way to produce complete organic fertilizer in quantities to relieve this serious menace to our health. In my books there is no end of proof and evidence covering fifty years of research which it but partly covers, but it gives sufficient data to satisfy any unbiased reader that the new era for world-wide improvement in soils and health is here for all to enjoy. It shows crops so healthy that they completely cover the soil in all cases, world records in grain, prizes for quality of crops, all attesting to their health promoting qualities. To me it is strange that such knowledge is not widespread. There is ample evidence of its need amongst our medical fraternity and the colleges. The only reason that I was induced to write on the subject was to present it for the welfare of sufferers who should be relieved, and to show that there are differences in organic wastes. Some are harmful and cannot produce good food because they carry wastes from wild molds, yeasts, and pathogenic bacteria, just as toadstools are poisonous in the mold world and mushrooms are healthful in the same mold or fungus world. A complete organic fertilizer is quite different to an ordinary compost, and while the compost may produce a large crop is no proof that the crop will have its total needs of vitamins, proteins, hormones, minerals, and essences sufficient to promote good health. Only a complete organic fertilizer can assure this quality. Even then it must be made under proper control to assure its purity. That is why I had a college put up a digestor to train students, so that there would be competent help to insure a good quality from plants making organic fertilizer for the future.

This work is done quite differently to the manufacture of acid chemical fertilizers. Such fertilizers carry three main elements, nitrogen, phosphorus and potash. Some companies try to add some of the trace elements. They can't add others because they are incompatible and would be ruined by combination. It would be useless adding valuable bacteria because they would be rapidly destroyed in such concentrated chemicals, so no mention is made of them. Yet they maintain fertility, as they have for centuries, in all uncultivated soils. They produce abundant food for all the great forests of the world. They supply the expensive nitrogen, also carbon, oxygen, and hydrogen to meet the crops' needs. Farmers have to buy their nitrogen bacteria separately as they cannot receive them with their fertilizer, nor can they drill them into the ground with the fertilizer.

When I introduced these bacteria in pure culture in 1905 very little was known of their use, and it took some persuasion for farmers to buy an invisible lot of bacteria, but the results were so outstanding that within five years over 90% of seed dealers sold our bacterial cultures to farmers throughout the country. Now millions of farmers use them to improve their legume crops. They still have to buy them separately to treat their seed and soil. Soon they will obtain them, and other groups of more valuable bacteria, all in one package with all the fertilizing elements complete. It will not only carry the bacteria, but increase their growth by supplying them with the organic food that all bacteria require. These hosts of beneficial bacteria will help to make foods tied up in an insoluble form in the land again available, thus enriching the soil at no additional cost to the farmer.

The organic fertilizer will not acidulate the land but by the oxidation produced by bacterial action bring about a more favorable reaction in soil for increased crop production. These beneficial bacteria in such a favorable organic medium grow in such vast numbers that they fight off the poisonous germs in soil by robbing and spoiling their acid foods. In my book this is well shown by agricultural authorities of the several provinces of France and Morocco where my complete organic fertilizer is made and used in an increasing volume. The same applies in Japan and in reports from Central America where large tonnage of organic fertilizer is made. Other pictures show crops grown here and abroad superior in vigor and production than by any other means. Best of all, it gives proof of the advantage in better health of crop, animal and man. It is not just my story but it includes supporting evidence of a university, its scientists, and others equally renowned who give testimony of the reduction of disease in animals and mankind when good soil provides complete nutrition. They discovered this remarkable fact by analyzing soil and the blood of those diseased. They found that the soil and the blood lacked some essential factors. They then supplied the patients with the missing factors, and all showed improvement, and most of them made complete recoveries within five months. So that in the future if we provide these essential factors to the sick soil we will not have these degenerative diseases, and those who have them can eat their way to improved health. I have for many years lectured and written of this prospect. Now it is not a prospect, it is an accomplished fact, verified by able men in hundreds of successful recoveries of chronic diseases beyond the help of drugs or antibiotics.

Many years ago, realizing this coming crisis in soil, I made provison to remedy the evil. I made a machine like a huge incinerator with moving parts to propel waste organic bacterial food slowly through it, with provision to supply graduated air for their growth. I called it a continuous flow digestor. It is the best means to dispose of city and country wastes and, at the same time, make a life-preserving complete plant food. I printed pictures of these digestors which have worked for many years, and exhibits of the splendid rich food the organic fertilizer has produced. There are many pages in the book giving priceless information on the whole subject of improved health through natural means by complete nutrition. I could have filled three books with the grateful letters and pictures I have received from people benefited by this new knowledge. There is a great deal about the sterling qualities of the mineral factors in maintaining vigor to old age.

I have two digestors making organic fertilizer at the laboratory and a small lucite model. I exhibited the model on the 10th of March over the Dumont Television network in Philadelphia, Pa. It aroused great interest, and we were awarded first prize. We have been invited to again show it working with garbage converting it into a black oxidized odorless humus on the 13th of April at 9:00 p.m. over the Dumont Television national hook up. The material can be seen passing out from the digestor by action of the motor moving the material through and out of it. It is the only rapid process in the world. It was especially constructed to do a most valuable service, to make superior plant food to improve crop health and that of its users.

This week we have visitors from England, France, and Italy to see our digestors working. Visitors have already called from Australia, where two large digestors are now converting city garbage into valuable organic fertilizers. Others came from France, Morocco, Cuba, and Mexico, Dominican Republic, Canada, and also from Florida, Chicago, Iowa, Nevada, Nebraska, Mississippi, New York, New Jersey, Kansas, Oregon, Washington, California, and Missouri. All of them to gain evidence of the use of digestors. Now that we can again purchase steel quite a number of digestors are ordered and being built. It requires very little labor to operate a digestor. Motors do most of the work because the bacteria produce the heat and break up the chunks of waste fed to them, and machinery puts it in the bag. All of this is well illustrated in pictures in the books. You will find in them many answers to ill health and how it was brought about and how to avoid it. You can get a new concept of life and pleasant living. It shows the way to a more pleasant and abundant life in a world-wide aspect.

If my life has contributed in a small degree to help this development then I have not lived in vain. I felt that as I had spent the better part of my life in this research that now I should write about it, since its soundness was no longer under a cloud of skepticism or doubt. All we now need is students to carry on and supply the world with its needs. I hope that my two books and the illustrations have been sufficient to show that the ground work has been laid for a world-wide improvement in our conservation or organic waste, our disposal systems, and our health and well being.

When General Douglas MacArthur was administering Japan he came across the records of our work in making fertilizer through Joske Komori, my agent in Japan. He asked his Aide-de-camp, Captain McGimpsey, to write us to say that the MacArthur team would assist us in every way to speedily re-establish making the fertilizer in Japan.

Captain Hart even saw Japanese fertilizer companies to speed up resumption of the work, finally sending a Japanese to see me about it. Despite General MacArthur's busy life in Japan his interest was in aiding his country as well as Japan. Our surprise and admiration of the man was unbounded. A sample letter will show how deep and sincere was the great General Administrator.

We also include excerpts from a few letters as references in relation to other views of this new science.

UNITED NATIONS
HEALTH AND WELFARE DEPT.

IRENE BALINSKA
17 EAST 70TH STREET
NEW YORK 21, N.Y.

April 9, 1951

Dear Dr. Earp-Thomas,

I have just read an article about the relation of ailments to the formation of cancer in a popular French magazine called "Constellation."

It is about the results obtained by a Dr. J. Kreitz in Vienna who is convinced the proper alimentation is absolutely essential as a factor in combating cancer. This will not surprise you naturally. But I thought you might like to know about this Dr. Kreitz. And if you had not heard about his work, drop me a line and I shall send you a rough translation of this interesting article—unless you can read it in the original—namely, French. The article is written by a Dr. F. M. Delacroit.

I am sure that it will not surprise you either that, taking what I call your "green powder" regularly throughout the winter has kept me "green" and hale without colds and/or catching the grippe everybody else seems to have had.

I meant to write you many times; reading about Dr. Kreitz finally catalyzed this intention into something more tangible.

With my best greetings to you and your family.

(Signed) I. B. M.D.
United Nations Health
and Welfare Dept.

COLUMBIA UNIVERSITY
IN THE CITY OF NEW YORK
SCHOOL OF BUSINESS

Swarthmore, Pa.
3/31/45

Dr. Earp-Thomas
Bloomfield Laboratories
Bloomfield, N. J.

I have heard most convincing accounts of the virtues of your preparation for improvement of health, especially arthritis.

Please do the following:

(a) Send me your descriptive material.

Yours,
(Signed) J. Russell Smith, Sc.D

NEW YORK UNIVERSITY
SCHOOL OF EDUCATION
WASHINGTON SQUARE
NEW YORK 3, N. Y.

Dept. of Physical Education

August 8, 1945

Dr. Earp-Thomas
Bloomfield Laboratories
Bloomfield, N. J.

Dear Dr. Earp-Thomas:

I believe that you have succeeded in developing a truly remarkable combination of food substances. You will be interested to hear that the arthritis has definitely improved. There is no mistaking the improvement.

What should be done now is to have carefully controlled laboratory experiments on animals conducted. I should like to see an experiment with white rats, using the ordinary type of American diet with both the control and experimental groups but feeding in addition an appropriate amount of this food combination, which you have tentatively called —— to the experimental group. What I would like to find out is whether or not the longevity of the experimental group would be increased by the use of this substance. I would also like to see controlled experiments with the use of this food combination for various types of disorders.

Very truly yours,
(Signed) M. M. H.
Assistant Professor
of Education

NEW YORK EXPERIMENT STATION
GENEVA, NEW YORK

September 5, 1914

To Whom It May Concern:

The secret process of the Earp-Thomas Company are all they are claimed to be. I went even farther in the tests by removing the plugs from the tubes of the growing plants, and there was no contamination of the jelly in the tubes of the growing plants.

(Signed) J. H. C.
Bacteriologist
Geneva Experiment Station
New York

NUTRITIONAL RESEARCH FOR
BETTER COMMUNITY HEALTH
M. O. GARTEN, D. O.
AUTHOR, WORLD TRAVELER, LECTURER
4103 PARK BOULEVARD
SAN DIEGO 3, CALIFORNIA
Phone Jackson 0911
August 6, 1951

Dr. G. H. Earp-Thomas, Director
Earp Laboratories
Hampton, New Jersey

Dear Doctor Earp-Thomas:

Have just completed reading your article in the July Herald of Health, and found it as being one of the most important writings of our times.

In India I visited with the Hunza Kutes of Hunza, people that live far beyond the 100-year span without any disease.

I am leaving September 20 for Argentina where I intend to spend some time with Dr. Raffo of the Argentine Cancer Research Hospital.

Sincerely yours,
(Signed) M. O. Garten, D.C.

Address:
Jacumba Hot Springs, Jacumba, Calif.

DR. H. R. C. F.
1819 BROADWAY, ROOM 805
NEW YORK, N. Y.
February 17, 1942

Earp Laboratories
Bloomfield, N. J.

Dear Dr. Thomas:

Last night I heard with the greatest pleasure your speech about "Mineral Deficiencies and Disease." I am interested in your —— for my patients.

Kindly give me further instructions concerning price, etc., and how do you sell this to the physicians.

Expecting your kindly answer.

Sincerely,
(Signed) Dr. H. R. C. F.
Dr. MAURICE H. KOWAN
Physician and Surgeon

LA CROSSE, VIRGINIA

January 16th, 1925

Dear Sirs:

Last July I wrote you about a child that I was going to put on your ——. This child had been in poor health for some time. Had about a year ago consulted a child specialist in Richmond, but had not improved any.

The child was 4½ years old when I gave him the —— and weighed 33½ lbs. I had him weighed when he had finished taking the bottle and he weighed 37 lbs. 2 oz.

I have not been to see the child as he lives six miles out in the country, but his father says that he never saw such improvement in any one.

Why I am writing you now is I suffer a good deal with gas on my stomach and I want you to send me a bottle of ——, I am going to take it myself.

Yours very truly,

(Signed) W. W. W., M.D.

105 EAST 10TH STREET
NEW YORK CITY

Bloomfield Laboratories
Bloomfield, New Jersey

Gentlemen:

I am getting wonderful results with your ——. I have had several cases of impotence due to intestinal toxemia covering a period of two years and the results were very prompt.

I never thought so many conditions were due to intestinal toxemia.

Yours respectfully,

(Signed) J. A. P., M.D.

ATLANTA BILTMORE
THE SOUTH'S SUPREME HOTEL
ATLANTA, GA.

Room 107, July 24, 1944

Dear Dr. Thomas:

My neighbor across the hall has told me of your wonderful

minerals and the marvelous results she and others have obtained from the use of them.

I am referring to Mrs. E. H. C., formerly of Washington, D.C. —— I was thrilled when I heard of your work. More power to you.

Sincerely,
(Signed) Miss L. V.

PALATKA, FLORIDA
S. L. F., INC.

April 8, 1953

Bloomfield Laboratories Corp.
Hampton, New Jersey

Dear Dr. Earp-Thomas:

We are following your valuable articles in the —— magazine. We deeply appreciate all that you have to say.

Your trace minerals have definitely proven a great help in treating cerebral palsy children. In our state some groups are still searching for ways and means to speed up their recovery, but with no measured success.

Just as you have said in your articles, that most of the mentally retarded and other exceptional children are the way they are, is because their mothers lacked the necessary minerals in their blood streams.

Sincerely yours,
(Mrs.) H. W. J.

New York, April 14, 1953

Doctor Earp-Thomas,
Earp Laboratories,
Hampton, New Jersey

Dear Doctor Earp-Thomas:

I am very happy to have come to the United States in order to get acquainted with you.

Indeed it is for me a great honor to have shaken the hand of a true scientist whose researches have had so important results either in improving human health and life or in renovating, with the help of your bacterias, the soil throughout the world, soils which have been wearing out for a great many years. And I do not overlook your products for bringing men and animals back to health.

For the past several months I was suffering from a back ache. I will tell you that 3 days after having taken your two products the pains disappeared. Really, it is a miracle.

I understand but too well why in general the scientific world bars the road to progress.

Unfortunately, my stay in the United States has been too short, but your cordial welcome was a delight to me.

I was able to see your Digestor in operation and also the one that Mr. Haldane is now building and which will be completed within a few days. I realize the scope of your efforts in surmounting the difficulties that you met before securing a perfect result.

I have just returned from Canada, where I saw another Digestor (CRANE—EARP-THOMAS), an apparatus that also runs satisfactorily.

Dear Doctor Earp-Thomas, I realize that official titles and degrees do not mean anything. Results alone count:

MOZART was a master.

RODIN was a master, yet neither had any title or degree.

Please remember me to Mrs. Earp-Thomas and your family, and receive my thanks for your kind assistance.

Very sincerely yours,
Armand G. Daudier

In an address given March 10, 1953 at the 18th annual conference of the National Farm Chemurgic Council of St. Louis, Mo., an eminent Bio-Chemist, Dr. H. G. P. gave an important speech entitled, "Returning Wastes to the Soil."

He mentioned the use of special bacteria used for spreading conversion of wastes in an Earp Thomas Continuous Flow Digestor, and other composting systems. All 12 pages of his article stress the essential need of organic matter to rejuvenate the soil to produce larger and better vegetation. He had pluck enough to be the first scientist to state that there was evidence to show that a good soil showed prospects of ameliorating diseases.

He says "It has been reported from time to time that crops grown in robust soil seems to be more resistant to plant diseases. We might borrow a page from the notebooks of animal nutritionists who have found that healthy rapid growth minimizes infections and susceptibility to disease."

His statement may give courage to other scientists to follow suit. Sorry I can't spare space for more of his excellent article. It should be in every farmer's hands. Dr. Nolfi spares no words to show that good organic soil produces food that cures most of her patients at her sanitorium in Denmark. Her instinct to produce good nutritious food from organic soil made her a pioneer in Organic Therapy.

God's World

The windswept hills lie robed in red,
Where fiery leaflets fell, and
Moonbeams dance a fantasy
On brook and wooded dell.

Night creatures scurry through the brush,
Intent to reach their lair.
The birds of flight are lost tonight,
The world retreats from care.

The scent of woodsmoke from below,
The distant church bells' toll,
Seals closed the mantle of the night
O'er man and his fettered role.

Within man's soul and limited frame,
The omniscient light grows bright
Received by some who know not Truth,
As the beauty of the night.

Man's conscious state sees not the light
That feeds his very soul
Despite his loss in his earthly trek
And the "lack of a cosmic role."

The ego and power, that men proclaim,
Draws tight the blinds of night
Against the ever awakening dawn
Of God's Eternal Light.

Look out and see God's world sublime,
Its radiance of many a hue
Look in and know the world divine
That came to those who knew!

Know God's World as not a sphere,
But the realms of space unknown.
Where dwells the light of love for all,
And God's great works are sown.

C. B. Brailey

fromouterspaceto00meng
fromouterspaceto00meng
fromouterspaceto00meng

CPSIA information can be obtained
at www.ICGtesting.com
Printed in the USA
LVHW021520230423
745117LV00002B/150